THINKING IN NUMBERS

Also by Daniel Tammet

Born on a Blue Day

Embracing the Wide Sky

THINKING IN NUMBERS

On Life, Love, Meaning, and Math

DANIEL TAMMET

Little, Brown and Company
New York Boston London

Little, Brown and Company
Hachette Book Group
237 Park Avenue, New York, NY 10017
littlebrown.com

First North American Edition, July 2013
Originally published in Great Britain by Hodder & Stoughton, August 2012

Little, Brown and Company is a division of Hachette Book Group, Inc.
The Little, Brown name and logo are trademarks of Hachette Book Group, Inc.

ISBN 978-0-316-18737-4
LCCN 2013935728

10 9 8 7 6 5 4 3 2 1

RRD-C

Printed in the United States of America

To see everything, the Master's eye is best of all,
As for me, I would add, so is the Lover's eye.

— Caius Julius Phaedrus

Like all great rationalists you believed in things that were twice as incredible as theology.

— Halldór Laxness,
Under the Glacier

Chess is life.

— Bobby Fischer

CONTENTS

PREFACE

Every afternoon, seven summers ago, I sat at my kitchen table in the south of England and worked on a book. Its name was *Born on a Blue Day.* The keys on my computer registered hundreds of thousands of impressions. Typing out the story of my formative years, I realized how many choices make up a single life. Every sentence or paragraph confided some decision I or someone else—a parent, teacher, or friend—had taken, or not taken. Naturally I was my own first reader, and it is no exaggeration to say that in writing, then reading the book, the course of my life was inexorably changed.

The year before that summer, I had traveled to the Center for Brain Studies in California, where the neurologists probed me with a battery of tests. It took me back to early days in a London hospital when, surveying my brain for seizure activity, the doctors had fixed me up to an encephalogram machine. Attached wires had streamed down and around my little head, until it resembled something hauled up out of the deep, like angler's swag.

These scientists wore tans and white smiles. They gave me sums to solve, and long sequences of numbers to learn by heart. Newer tools measured my pulse and my breathing as I thought. I submitted to all these experiments with a burning curiosity; it felt exciting to learn the secret of my childhood.

My autobiography opens with their diagnosis. My difference finally had a name. Until then it had run the whole gamut of inventive aliases: painfully shy, hypersensitive, cack-handed (in my father's characteristically colorful words). According to the scientists, I had high-functioning autistic savant syndrome: the connections in my brain, since birth, had formed unusual circuits. Back home in England I began to write, with their encouragement, producing pages that in the end found favor with a London editor.

To this day, readers both of the first book and of my second, *Embracing the Wide Sky,* continue to send me their messages. They wonder how it must be to perceive words and numbers in different colors, shapes, and textures. They try to picture solving a sum in their mind using these multidimensional colored shapes. They seek the same beauty and emotion that I find in both a poem and a prime number. What can I tell them?

Imagine.

Close your eyes and imagine a space without limits, or the infinitesimal events that can stir up a country's revolution. Imagine how the perfect game of chess might start and end: a win for white, or black, or a draw? Imagine numbers so vast that they exceed every atom in the universe, counting with eleven or twelve fingers instead of ten, reading a single book in an infinite number of ways.

Such imagination belongs to everyone. It even possesses

its own science: mathematics. Ricardo Nemirovsky and Francesca Ferrara, who specialize in the study of mathematical cognition, write that "like literary fiction, mathematical imagination entertains pure possibilities." This is the distillation of what I take to be interesting and important about the way in which mathematics informs our imaginative life. Often we are barely aware of it, but the play between numerical concepts saturates the way we experience the world.

This new book, a collection of twenty-five essays on the "math of life," entertains pure possibilities. According to the definition offered by Nemirovsky and Ferrara, "pure" here means something immune to prior experience or expectation. The fact that we have never read an endless book, or counted to infinity (and beyond!) or made contact with an extraterrestrial civilization (all subjects of essays in the book) should not prevent us from wondering: *what if?*

Inevitably, my choice of subjects has been wholly personal and therefore eclectic. There are some autobiographical elements but the emphasis throughout is outward looking. Several of the pieces are biographical, prompted by imagining a young Shakespeare's first arithmetic lessons in zero—a new idea in sixteenth-century schools—or the calendar created for a Sultan by the poet and mathematician Omar Khayyám. Others take the reader around the globe and back in time, with essays inspired by the snows of Quebec, sheep counting in Iceland, and the debates of ancient Greece that facilitated the development of the Western mathematical imagination.

Literature adds a further dimension to the exploration of those pure possibilities. As Nemirovsky and Ferrara suggest, there are numerous similarities in the patterns of thinking and creating shared by writers and mathematicians (two vocations often considered incomparable). In the "Poetry of the Primes," for example, I explore the way in which certain poems and number theory coincide. At the risk of disappointing fans of "mathematically constructed" novels, I admit this book has been written without once mentioning the name "Perec."

The following pages attest to the changes in my perspective over the seven years since that summer in southern England. Travels through many countries in pursuit of my books as they go from language to language, accumulating accents, have contributed much to my understanding. Exploring the many links between mathematics and fiction has been another spur. Today, I live in the heart of Paris. I write full-time. Every day I sit at a table and ask myself: *what if?*

THINKING IN NUMBERS

One

FAMILY VALUES

In a smallish London suburb where nothing much ever happened, my family gradually became the talk of the town. Throughout my teens, wherever I went, I would always hear the same question, "How many brothers and sisters do you have?"

The answer, I understood, was already common knowledge. It had passed into the town's body of folklore, exchanged between the residents like a good yarn.

Ever patient, I would dutifully reply, "Five sisters, and three brothers."

These few words never failed to elicit a visible reaction from the listener: brows would furrow, eyes would roll, lips would smile. "Nine children!" they would exclaim, as if they had never imagined that families could come in such sizes.

It was much the same story in school. *"J'ai une grande famille"* was among the first phrases I learned to say in Monsieur Oiseau's class. From my fellow students, many of whom were single sons or daughters, the sight of us siblings together attracted comments that ranged all the way from faint disdain to outright awe. Our peculiar fame became such that for a time it outdid every other in the town: the one-handed grocer, the enormously obese Indian girl, a neighbor's singing dog, all found themselves temporarily displaced in the local gossip. Effaced as individuals, my brothers, sisters, and I existed only in number. The quality of our quantity became something we could not escape. It preceded us everywhere: even in French, whose adjectives almost always follow the noun (but not when it comes to *une grande famille*).

With so many siblings to keep an eye on, it is perhaps little wonder that I developed a knack for numbers. From my family I learned that numbers belong to life. The majority of my math acumen came not from books but from regular observations and day-to-day interactions. Numerical patterns, I realized, were the matter of our world. To give an example, we nine children embodied the decimal system of numbers: zero (whenever we were all absent from a place) through to nine. Our behavior even bore some resemblance to the arithmetical: over angry words, we sometimes divided; shifting alliances between my brothers and sisters combined and recombined them into new equations.

We are, my brothers, sisters, and I, in the language of mathematicians, a "set" consisting of nine members. A mathematician would write:

S = {Daniel, Lee, Claire, Steven, Paul, Maria, Natasha, Anna, Shelley}

Put another way, we belong to the category of things that people refer to when they use the number nine. Other sets of this kind include the planets in our solar system (at least, until Pluto's recent demotion to the status of a non-planet), the squares in a game of x's and o's, the players in a baseball team, the muses of Greek mythology, and the Justices of the U.S. Supreme Court. With a little thought, it is possible to come up with others, including:

> {February, March, April, May, August, September, October, November, December} where S = the months of the year not beginning with the letter J.
>
> {5, 6, 7, 8, 9, 10, Jack, Queen, King} where S = in poker, the possible high cards in a straight flush.
>
> {1, 4, 9, 16, 25, 36, 49, 64, 81} where S = the square numbers between 1 and 99.
>
> {3, 5, 7, 11, 13, 17, 19, 23, 29} where S = the odd primes below 30.

There are nine of these examples of sets containing nine members, so taken together they provide us with a further instance of just such a set.

Like colors, the commonest numbers give character, form, and dimension to our world. Of the most frequent—zero and one—we might say that they are like black and white, with the other primary colors—red, blue, and yellow—akin to two, three, and four. Nine, then, might be a sort of cobalt or indigo: in a painting it would contribute shading, rather than shape. We expect to come across samples of nine as we might samples of a color like indigo—only occasionally, and in small and subtle ways. Thus a family of nine children surprises as much as a man or woman with cobalt-colored hair.

I would like to suggest another reason for the surprise of my town's residents. I have alluded to the various and alternating combinations and recombinations between my siblings. In how many ways can any set of nine members divide and combine? Put another way, how large is the set of all subsets?

{Daniel} . . . {Daniel, Lee} . . . {Lee, Claire, Steven} . . . {Paul} . . . {Lee, Steven, Maria, Shelley} . . . {Claire, Natasha} . . . {Anna} . . .

Fortunately, this type of calculation is very familiar to mathematicians. As it turns out, we need only to multiply the number two by itself, as many times as there are members in the set. So, for a set consisting of nine members the answer to our question amounts to: $2 \times 2 \times 2 \times 2 \times 2 \times 2 \times 2 \times 2 \times 2 = 512$.

This means that there existed in my hometown, at any given place and time, 512 different ways to spot us in one or another combination. 512! It becomes clearer why we attracted so much attention. To the other residents, it really must have seemed that we were legion.

Here is another way to think about the calculation that I set out above. Take any site in the town at random, say a classroom or the municipal swimming pool. The first "2" in the calculation indicates the odds of my being present there at a particular moment (one in two — I am either there, or I am not). The same goes for each of my siblings, which is why two is multiplied by itself a total of nine times.

In precisely one of the possible combinations, every sibling is absent (just as in one of the combinations we are all present). Strange as it may sound, we can even define those sets containing no objects. Mathematicians call them an "empty set." Where sets of nine members embody everything we can think of, touch, or point to when we use the number nine, empty sets are all those that are represented by the value zero. So while a Christmas reunion in my hometown can bring together as many of us as there are Justices on the U.S. Supreme Court, a trip to the moon will unite only as many of us as there are pink elephants, four-sided circles, or people who have swum the breadth of the Atlantic Ocean.

Our mind uses sets when we think and when we perceive just as much as when we count. Our possible thoughts and perceptions about these sets can range almost without limit. Fascinated by the different cultural subdivisions and categories of an infinitely complex world, the Argentine writer Jorge Luis Borges offers a mischievously tongue-in-cheek illustration in his fictional Chinese encyclopedia entitled *The Celestial Emporium of Benevolent Knowledge:*

> Animals are classified as follows: (a) those that belong to the Emperor; (b) embalmed ones; (c) those that are trained; (d) suckling pigs; (e) mermaids; (f) fabulous ones; (g) stray dogs; (h) those that are included in this

> classification; (i) those that tremble as if they were mad; (j) innumerable ones; (k) those drawn with a very fine camel's-hair brush; (l) et cetera; (m) those that have just broken the flower vase; (n) those that at a distance resemble flies.

Never one to forgo humor in his texts, Borges here also makes several thought-provoking points. First, though a set as familiar to our understanding as that of "animals" implies containment and comprehension, the sheer number of its possible subsets actually swells toward infinity. With their handful of generic labels ("mammal," "reptile," "amphibious," etc.), standard taxonomies conceal this fact. To say, for example, that a flea is tiny, parasitic, and a champion jumper is only to begin to scratch the surface of all its various aspects.

Second, defining a set owes more to art than it does to science. Faced with the problem of a near endless number of potential categories, we are inclined to choose from a few—those most tried and tested within our particular culture. Western descriptions of the set of all elephants privilege subsets like "those that are very large," and "those possessing tusks," and even "those possessing an excellent memory," while excluding other equally legitimate possibilities such as Borges's "those that at a distance resemble flies," or the Hindu "those that are considered lucky."

Memory is a further example of the privileging of certain subsets (of experience) over others, in how we talk and think about a category of things. Asked about his birthday, a man might at once recall the messy slice of chocolate cake that he devoured, his wife's enthusiastic embrace, and the pair of fluorescent green socks that his mother presented to him. At the same time, many hundreds, or thousands, of other details likewise composed his special day, from the mundane (the crumbs from his morning toast that he brushed out of his lap) to the peculiar (a sudden hailstorm on the mid-July afternoon that lasted several minutes). Most of these subsets, though, would have completely slipped his mind.

Returning to Borges's list of subsets of animals, several of the categories pose paradoxes. Take, for example, the subset (j): "innumerable ones." How can any subset of something—even if it is imaginary, like Borges's animals—be infinite in size? How can a part of any collection not be smaller than the whole?

Borges's taxonomy is clearly inspired by the work of Georg Cantor, a nineteenth-century German mathematician whose important discoveries in the study of infinity provide us with an answer to this paradox.

Cantor showed, among other things, that parts of a collection (subsets) as great as the whole (set) really do exist. Counting, he observed, involves matching the members of one set to another. "Two sets *A* and *B* have the same number of members if and only if there is a perfect one-to-one correspondence between them." So, by matching each of my siblings and myself to a player on a baseball team, or to a month of the year not beginning with the letter J, I am able to conclude that each of these sets is equivalent, all containing precisely nine members.

Next came Cantor's great mental leap: in the same manner, he compared the set of all natural numbers (1, 2, 3, 4, 5 . . .) with each of its subsets such as the even numbers (2, 4, 6, 8, 10 . . .), odd numbers (1, 3, 5, 7, 9 . . .), and the primes (2, 3, 5, 7, 11 . . .). Like the perfect matches between each of the baseball team players and my siblings and me, Cantor found that for each natural number he could uniquely assign an even, an odd, and a prime number. Incredibly, he concluded, there are as "many" even (or odd, or prime) numbers as all the numbers combined.

Reading Borges invites me to consider the wealth of possible subsets into which my family "set" could be classified, far beyond those that simply point to multiplicity. All grown up today, some of my siblings have children of their own. Others have moved far away, to the warmer and more interesting places from where postcards come. The opportunities for us all to get together are rare, which is a great pity. Naturally I am biased, but I love my family. There is a lot of my family to love. But size ceased long ago to be our defining characteristic. We see ourselves in other ways: those that are studious, those that prefer coffee to tea, those that have never planted a flower, those that still laugh in their sleep . . .

Like works of literature, mathematical ideas help expand our circle of empathy, liberating us from the tyranny of a single, parochial point of view. Numbers, properly considered, make us better people.

Two

ETERNITY IN AN HOUR

Once upon a time I was a child who loved to read fairy tales. Among my favorites was "The Magic Porridge Pot" by the Brothers Grimm. A poor, good-hearted girl receives from a sorceress a little pot capable of spontaneously concocting as much sweet porridge as the girl and her mother can eat. One day, after eating her fill, the mother's mind goes blank and she forgets the magic words "Stop, little pot."

> So it went on cooking and the porridge rose over the edge, and still it cooked on until the kitchen and whole house were full, and then the next house, and then the whole street, just as if it wanted to satisfy the hunger of the whole world.

Only the daughter's return home, and the requisite utterance, finally brings the gooey avalanche to a belated halt.

The Brothers Grimm introduced me to the mystery of infinity. How could so much porridge emerge from so small a pot? It got me thinking. A single flake of porridge was awfully slight. Tip it inside a bowl and one would probably not even spot it for the spoon. The same held for a drop of milk, or a grain of sugar.

What if, I wondered, a magical pot distributed these tiny flakes of porridge and drops of milk and grains of sugar in its own special way? In such a way that each flake and each drop and each grain had its own position in the pot, released from the necessity of touching. I imagined five, ten, fifty, one hundred, one thousand flakes and drops and grains, each indifferent to the next, suspended here and there throughout the curved space like stars. More porridge flakes, more drops of milk, more grains of sugar are added one after another to this evolving constellation, forming microscopic Big Dippers and minuscule Great Bears. Say we reach the ten thousand four hundred and seventy-third flake of porridge. Where do we include it? And here my child's mind imagined all the tiny gaps—thousands of them—between every flake of porridge and drop of milk and grain of sugar. For every minute addition, further tiny gaps would continue to be made. So long as the pot magically prevented any contact between them, every new flake (and drop and granule) would be sure to find its place.

Hans Christian Andersen's "The Princess and the Pea" similarly sent my mind spinning toward the infinite, but this time, an infinity of fractions. One night, a young woman claiming to be a princess knocks at the door of a castle. Outside, a storm is blowing and the pelting rain musses her clothes and turns her golden hair black. So sorry a sight is she that the queen of the castle doubts her story of high birth. To test the young woman's claim, the queen decides to place a pea beneath the bedding on which the woman will sleep for the night (princesses being most delicate creatures!). Her bed is piled to a height of twenty mattresses. But in the morning the woman admits to having hardly slept a wink.

The thought of all those tottering mattresses kept me up long past my own bedtime. By my calculation, a second mattress would double the distance between the princess's back and the offending pea. The tough little legume would therefore be only half as prominent as before. Another mattress reduces the pea's prominence to one-third. But if the young princess's body is sensitive enough to detect one-half of a pea (under two mattresses) or one-third of a pea (under three mattresses), why would it not also be sensitive enough to detect one-twentieth? In fact, possessing limitless sensitivity (this is a fairy tale after all), not even one-hundredth, or one-thousandth or one-millionth of a pea could be tolerably borne.

Which brings us back to the Brothers Grimm and their tale of porridge. For the princess, even a pea felt infinitely big; for the poor daughter and her mother, even an avalanche of porridge reduced to the infinitesimally small.

"You have too much imagination," my father said when I shared these thoughts with him. "You always have your nose in some book." My father kept a pile of paperbacks and regularly bought the weekend papers, but he was never a particularly enthusiastic reader. "Get outdoors more—there's no good being cooped up in here."

Hide-and-seek in the park with my brothers and sisters lasted all of ten minutes. The swings held my attention for about as long. We walked the perimeter of the lake and threw breadcrumbs out onto the grimy water. Even the ducks looked bored.

Games in the yard offered greater entertainment. We fought wars, cast spells, and traveled back in time. In a cardboard box we sailed along the Nile; with a bedsheet we pitched a tent in the hills of Rome. At other times, I would simply stroll the local streets to my heart's content, dreaming up all manner of new adventures and imaginary expeditions.

Returning one day from China, I heard the low grumble of an approaching storm and fled for cover inside the municipal library. Everyone knew me there; I was one of their regulars. Corridors of books pullulated around them, centuries of learning tiled the walls, and I brushed my fingertips along the seemingly endless shelves as I walked.

My favorite section brimmed with dictionaries and encyclopedias: the building blocks of books. These seemed to promise (though of course they could not deliver) the sum of human knowledge: every fact and idea and word. This vast panoply of information was tamed by divisions—A–C, D–F, G–I—and every division subdivided in turn—Aa–Ad...Di–Do...Il–In. Many of these subdivisions also subdivided—Hai–Han...Una–Unf—and some among them subdivided yet further still—Inte–Intr. Where does a person start? And, perhaps more important, where should he stop? I usually left the choice to chance. At random I tugged an encyclopedia from the shelf and let its pages open where they may, and for the next hour I sat and read about *Bora Bora* and *borborygmi* and the *Borg* scale.

Lost in thought, I did not immediately notice the insistent *tap-tap* of approaching footsteps on the polished floor. They belonged to one of the senior librarians, a neighbor. He was tall (but then, to a child is not everyone tall?) and thin with a long head finished off by a few random sprigs of graying hair.

"I have a book for you," said the librarian. I craned upward a moment before taking the recommendation from his big hands. The cover, titled *The Borrowers,* wore a "Bookworms Club Monthly Selection" sticker. I thanked him, less out of gratitude than the desire to end the sudden eclipse that darkened my table. But when I finally left the desk an hour later, the book left with me, checked out and tucked firmly beneath my arm.

It told the story of a tiny family that lived under the floorboards of a house. I tried to imagine what it would be like to live so small. In my mind's eye, I pictured the world as it continued to contract. The smaller I became, the bigger my surroundings grew. The familiar now became strange; the strange became familiar. All at once, a face of ears and eyes and hair becomes a pink expanse of shrub and grooves and heat. Even the tiniest fish becomes a whale. Specks of dust take flight like birds, swooping and wheeling above my head. I shrank until all that was familiar disappeared completely, until I could no longer tell a mound of laundry or a rocky mountain apart.

At my next visit to the library, I duly joined the Bookworms Club. The months were each twinned with a classic story, and some of the selections enchanted me more than others, but it was December's tale that truly seized my senses: *The Lion, the Witch, and the Wardrobe* by C. S. Lewis. Opening its pages, I followed Lucy, as she was sent with her siblings "away from London during the war because of the air raids . . . to the house of an old professor who lived in the heart of the country." It was "the sort of house that you never seem to come to the end of, and it was full of unexpected places."

With Lucy, I stepped into the large wardrobe in one of the otherwise bare spare rooms, tussled with its rows of dense and dust-fringed clothes as we fumbled our way with outstretched fingers toward the back. I, too, suddenly heard the crunch of snow beneath my shoes, and saw the fur coats give way all at once to the fir trees of this magical land, a wardrobe's depth away.

Narnia became one of my favorite places, and I visited it many times that winter. Repeated readings of the story would keep me in bright thoughts and images for many months.

One day, on the short walk home from school, it so happened that these images came to the front of my mind. The lampposts that lined the street reminded me of the lamppost I had read about in the story, the point in the landscape from which the children return to the warmth and mothballs of the professor's wardrobe.

It was midafternoon, but the electric lights were already

shining. Fluorescent haloes stood out in the darkening sky at equidistant points. I counted the time it took me to step with even paces from one lamppost to another. Eight seconds. Then I retreated, counting backward, and arrived at the same result. A few doors down, the lights came on in my parents' house; yellow rectangles glowed dimly between the red bricks. I watched them with only half a mind.

I was contemplating those eight seconds. To reach the next lamppost I had only to take so many steps. Before I reached there I would first have to arrive at the midway point. That would take me four seconds. But this observation implied that the remaining four seconds also contained a midway point. Six seconds from the start, I would land upon it. Two seconds would now separate me from my destination. Yet before I made it, another halfway point—a second later—would intervene. And here I felt my brain seethe hot under my woollen hat. For after seven seconds, the eighth and final second would likewise contain a halfway point of its own. Seven and one half seconds after starting off, the remaining half a second would also not elapse before I first passed its midway point in turn. After seven seconds and three-quarters, a stubborn quarter of a second of my journey would still await me. Going halfway through it would leave me an eighth of a second still to go. One sixteenth of a second would keep me from my lamppost, then 1/32 of a second, then 1/64, then 1/128, and so on. Fractions of fractions of fractions of a second would always distance me from the end.

Suddenly I could no longer depend on those eight seconds

to deliver me to my destination. Worse, I could no longer be sure that they would let me move one inch. Those same interminable fractions of seconds that I had observed toward the end of my journey applied equally to the start. Say my opening step took one second; this second, of course, contained a halfway point. And before I could cross this half of a second, I would first have to traverse its own midway point (the initial quarter second), and so on.

And yet, my legs disposed of all these fractions of seconds as they had always done. Adjusting the heavy schoolbag on my back, I walked the length between the lampposts and counted once again to eight. The word rang out defiantly into the cold crisp air. The silence that followed, however, was short-lived. "What are you doing standing outside in the cold and dark?" shouted my father from the yellow oblong of the open front door. "Come inside now."

I did not forget the infinity of fractions that lurked between the lampposts on my street. Day after day, I found myself slowing involuntarily to a crawl as I passed them, afraid perhaps of falling between the whole seconds into their interspersed gaps. What a sight I must have made, inching warily forward tiny step after tiny step with my round woolly head and the lumpy bag upon my back.

Numbers within numbers, and so tiny! I was amazed. These fractions of fractions of fractions of fractions of fractions went on forever. Add any of them to zero and they hardly registered at all. Add tens, hundreds, thousands, millions, billions of them to zero and the result is still almost exactly zero. Only infinitely many of these fractions could lead from zero to one, from nothing to something:

> 1/2 + 1/4 + 1/8 + 1/16 + 1/32 + 1/64 + 1/128 + 1/256 + 1/512 + 1/1024 . . . = 1.

One evening in the new year, my mother, very flustered, asked me to be on my best behavior. Guests—a rarity—were due any moment now, for dinner. My mother, it seemed, intended to repay some favor to the librarian's wife. "No funny questions," she said, "and no elbows on the table. And after the first hour, bed!"

The librarian and his wife arrived right on time with a bottle of wine that my parents never opened. With their backs to one another, they thrashed themselves out of their coats before sitting at the dining room table, side by side. The wife offered my mother a compliment about the checkered tablecloth. "Where did you buy it?" she asked, over her husband's sigh.

We ate my father's roast chicken and potatoes with peas and carrots, and as we ate the librarian talked. All eyes were on him. There were words on the weather, local politics, and all the nonsense that was interminably broadcast on TV. Beside him, his wife ate slowly, one-handed, while the other hand worried her thin black hair. At one point in her husband's monologue, she tried gently tapping his tightly bunched hand.

"What? What?"

"Nothing." Her fork promptly retired to her plate. She looked to be on the verge of tears.

Very much novices in the art of hospitality, my mother and father exchanged helpless glances. Plates were hurriedly collected, and bowls of ice cream served. A frosty atmosphere presently filled the room.

I thought of the infinitely many points that can divide the space between two human hearts.

Three

COUNTING TO FOUR IN ICELANDIC

Ask an Icelander what comes after three and he will answer, "Three what?" Ignore the warm blood of annoyance as it fills your cheeks, and suggest something, or better still, point. "Ah," our Icelander replies. Ruffled by the wind, the four sheep stare blankly at your index finger. *"Fjórar,"* he says at last.

However, when you take your phrase book—presumably one of those handy, rain-resistant brands—from your pocket and turn to the numbers page, you find, marked beside the numeral 4, *fjórir.* This is not a printing error, nor did you hear the Icelander wrong. Both words are correct; both words mean "four." This should give you your first inkling of the sophistication with which these people count.

I first heard Icelandic several years ago during a trip to Reykjavík. No phrase book in my pocket, thank God. I came with nothing more useful than a vague awareness of the shape and sounds of Old English, some secondary-school German, and plenty of curiosity. The curiosity had already seen service in France. Here in the North, too, I favored conversation over textbooks.

I hate textbooks. I hate how they shoehorn even the most incongruous words—like "cup" and "bookcase," or "pencil" and "ashtray"—onto the same page, and then call it "vocabulary." In a conversation, the language is always fluid, moving, and you have to move with it. You walk and talk and see where the words come from, and where they should go. It was in this way that I learned how to count like a Viking.

Icelanders, I learned, have highly refined discrimination for the smallest quantities. "Four" sheep differ in kind from "four," the abstract counting word. No farmer in Hveragerði would ever dream of counting sheep in the abstract. Nor, for that matter, would his wife or son or priest or neighbor. To list both words together, as in a textbook, would make no sense to them whatsoever.

This numerical diversity applies not only to sheep. Naturally enough, the woolly mammals feature little in town dwellers' talk. Like you and me, my friends in Reykjavík talk about birthdays and buses and pairs of jeans but, unlike in English, in Icelandic these things each require their own set of number words.

For example, a toddler who turns two is *tveggja* years old. And yet the pocket phrase book will inform you that "two" is *tveir.* Age, abstract as counting to our way of thinking, becomes in Icelandic a tangible phenomenon. Perhaps you too sense the difference: the word *tveggja* slows the voice, suggesting duration. We hear this possibly even more clearly in the word for a four-year-old: *fjögurra.* Interestingly, these sounds apply almost exclusively to the passage of years—the same words are hardly ever used to talk about months, days, or weeks. Clock time, on the other hand, renders the Icelander almost terse as a tick: the hour after one o'clock is *tvö*.

What about buses? Here numbers refer to identity rather than quantity. In Britain or America, we say something like, "the number three bus," turning the number into a name. Icelanders do something similar. Their most frequent buses are each known by a special number word. In Reykjavík, the number three bus is simply *þristur* (whereas to count to three the Icelander says *þrír*). *Fjarki* is how to say "four" when talking buses in Iceland.

A third example is pairs of something—whether jeans or shorts, socks or shoes. In this case, Icelanders consider "one" as being plural: *einar* pair of jeans, instead of the phrase book *einn*.

With time and practice, I have learned all these words, more words for the numbers one to four than has an English speaker to count all the way up to fifty. In English, I would suggest, numbers are considered more or less ethereal—as categories, not qualities. Not so the smallest numbers in Icelandic. It is as though each corresponded to a delicate nuance of color. Where the English word "red" is abstract, indifferent to its object, words like "crimson," "scarlet," and "burgundy" possess their own particular shade of meaning and application.

We can only speculate as to the reason why Icelanders stop at the number five (for which, like every number thereafter, a single word exists). According to psychologists, humans can count in flashes only up to quantities of four. We see three buttons on a shirt and say "three"; we glance at four books on a table and say "four." No conscious thought attends this process—it seems to us as effortless as the speech with which we pronounce the words. The same psychologists tell us that the smallest numbers loom largest in our minds. Asked to pick a number between one and fifty, we tend toward the shallow end of the scale (far fewer say "forty" than "fourteen"). It is one possible explanation for why only the commonest quantities feel real to us, why most numbers we accept only on the word of a teacher or textbook. Forty, to us, is but a vague notion; fourteen, on the other hand, is a quantity within our reach. Four, we recognize as something solid and definite. In Icelandic, you can give your baby the name "Four."

I do not know Chinese, but I have read that counting in this language rivals in its sophistication even that of the Icelanders. A shepherd in rural China says *sì zhī* when his flock numbers four, whereas a horseman—possessing the same quantity of horses—counts them as *sì pǐ*. This is because mounts are counted differently from other animals in Chinese. Domesticated animals, too. Asked how many cows he had milked that morning, a farmer would reply *sì tóu* (four). Fish are a further exception. *Sì tiáo* is how an angler would count his fourth catch of the day.

Unlike in Icelandic, in Chinese these fine distinctions apply to all quantities. What saves its speakers from endless trouble of recall is generalization. *Sì tiáo* means "four" when counting fish, but also trousers, roads, and rivers (and other long, slender, and flexible objects). A locksmith might enumerate his keys as *qī bǎ* (seven), but so too would a housewife apropos her seven knives, or a tailor when adding up his seven pairs of scissors (or other handy items). Imagine that, with his scissors, the tailor snips a sheet of fabric in two. He would say he has *liǎng zhāng* (two) sheets of fabric, using the same number word as he would for paper, paintings, tickets, blankets, and bedsheets. Now picture the tailor as he rolls the fabric into long inflexible tubes. This pair he counts as *liǎng juǎn* (two scrolls, or rolls of film, would be counted in the same way). Scrunching the sheets into balls, the tailor counts them as *liǎng tuán,* insofar as they resemble other pairs of round things.

When counting people, the Chinese start from *yī ge* (one), though for villagers and family members they begin *yī kǒu,* and *yī míng* for lawyers, politicians, and royals. Numbering a crowd thus depends on its composition. A hundred marchers would be counted as *yībǎi ge* if they consist, for example, of students, but as *yībǎi kǒu* if they hail from the villages.

So complex is this method of counting that, in some regions of China, the words for certain numbers have even taken on the varying properties of a dialect. *Wǔshí li,* for example, a standard Mandarin word meaning "fifty" (when counting small, round objects like grains of rice), sounds truly enormous to the speakers of southern Min, for whom it is used to count watermelons.

This profusion of Icelandic and Chinese words for the

purpose of counting appears to be an exception to the rule. Many of the world's tribal languages, in contrast, make do with only a handful of names for numbers. The Veddas, an indigenous people of Sri Lanka, are reported to have only words for the numbers one (*ekkamai*) and two (*dekkamai*). For larger quantities, they continue: *otameekai, otameekai, otameekai*... ("and one more, and one more, and one more..."). Another example is the Caquintes of Peru, who count one (*aparo*) and two (*mavite*). Three they call "it is another one"; four is "the one that follows it."

In Brazil, the Munduruku relay quantity by according an extra syllable to each new number: one is *pug,* two is *xep xep,* three is *ebapug,* and four is *edadipdip.* They count, understandably, no higher than five. This imitative method, while transparent, has clear limitations. Just imagine a number word as many syllables long as the quantity of trees leading to a food source! The seemingly endless chain of syllables would prove far too expensive to the tongue (not to say the listener's powers of concentration). It pains the head even to think about what it would be like to have to learn to recite the ten times tables in this way.

All this may sound almost incomprehensible for those brought up speaking languages that count to thousands, millions and beyond, but it does at least make the relationship between a quantity and its appointed word sound straightforward and conventional. Quite often, though, it is not. In many tribal languages, we find that the names for numbers are perfectly interchangeable, so that a word for "three" will also sometimes mean "two," and at other times "four" or "five." A word meaning "four" will have "three" and "five"—occasionally "six"—as synonyms.

Few circumstances within these communities require any greater numerical precision. Any number beyond their fingertips is superfluous to their traditional way of life. In many of these places there are, after all, no legal documents that require dates, no bureaucracies that levy taxes, no clocks or calendars, no lawyers or accountants, no banks or banknotes, no thermometers or weather reports, no schools, no books, no playing cards, no lines, no shoes (and hence, no shoe sizes), no shops, no bills, and no debts to settle. It would make as much sense to tell them, say, that a group of men amounts exactly to eleven, as for someone to inform us that this same group has precisely one hundred and ten fingers, and as many toes.

There is a tribe in the Amazon rain forest that knows nothing whatsoever of numbers. Their name is the *Pirahã* or the *Hi'aiti'ihi,* meaning "the straight ones." Surrounded by throngs of trees, their small clusters of huts lie on the banks of the Maici River. Tumbling gray rain breaks green on the lush foliage and long grass. Days there are continuously hot and humid, inducing a perpetual look of embarrassment on the faces of visiting missionaries and linguists. Children race naked around the village, while their mothers wear light dresses obtained by bartering with the Brazilian traders. From the same source, the men display colorful T-shirts, the flotsam of past political campaigns, exhorting the observer to vote Lula.

Manioc (a tough and bland tuber), fresh fish, and roasted

anteater sustain the population. The work of gathering food is divided along lines of sex. At first light, women leave the huts to tend the manioc plants and collect firewood, while the men go upriver or downriver to fish. They can spend the whole day there, bow and arrow in hand, watching the water. For want of means of storage, any catch is consumed quickly. The *Pirahã* apportion food in the following manner: members of the tribe haphazardly receive a generous serving until no more remains. Any who have not yet been served ask a neighbor, who has to share. This procedure only ends when everyone has eaten his fill.

The vast majority of what we know about the *Pirahã* is due to the work of Daniel Everett, a Californian linguist who has studied them at close quarters over a period of thirty years. With professional perseverance, he gradually heard their cacophonic ejections as comprehensible words and phrases, becoming in the process the first outsider to embrace the tribe's way of life.

To Everett's astonishment, the language he learned has no specific words for measuring time or quantity. Names for numbers like "one" or "two" are unheard of. Even the simplest numerical queries brought only confusion or indifference to the tribesmen's eyes. Of their children, parents are unable to say how many they have, though they remember all their names. Plans or schedules older than a single day have no purchase on the *Pirahã*'s minds. Bartering with foreign traders simply consists of handing over foraged nuts as payment until the trader says that the price has been met.

Nor do the *Pirahã* count with their bodies. Their fingers

never point or curl: when indicating some amount they simply hold their hand palm down, using the space between their hand and the ground to suggest the height of the pile that such a quantity could reach.

It seems the *Pirahã* make no distinction between a man and a group of men, between a bird and a flock of birds, between a grain of manioc flour and a sack of manioc flour. Everything is either small (*hói*) or big (*ogii*). A solitary macaw is a small flock; the flock, a big macaw. In his *Metaphysics,* Aristotle shows that counting requires some prior understanding of what "one" is. To count five, or ten or twenty-three birds, we must first identify one bird, an idea of "bird" that can apply to every possible kind. But such abstractions are entirely foreign to the tribe.

With abstraction, birds become numbers. Men and maniocs, too. We can look at a scene and say, "There are two men, three birds, and four maniocs" but also, "There are nine things" (summing two and three and four). The *Pirahã* do not think this way. They ask, "What are these things?" "Where are they?" "What do they do?" A bird flies, a man breathes, and a manioc plant grows. It is meaningless to try to bring them together. Man is a small world. The world is a big manioc.

It is little surprise to learn that the *Pirahã* perceive drawings and photos only with great difficulty. They hold a photograph sideways or upside down, not seeing what the image is meant to represent. Drawing a picture is no easier for them, not even a straight line. They cannot copy simple shapes with any fidelity. Quite possibly, they have no interest in doing so.

Instead their pencils (furnished by linguists or missionaries) produce only repeating circular marks on the researcher's sheet of paper, each mark a little different from the last.

Perhaps this also explains why the *Pirahã* tell no stories, possess no creation myths. Stories, at least as we understand them, have intervals: a beginning, a middle, and an end. When we tell a story, we recount: naming each interval is equivalent to numbering it. Yet the *Pirahã* talk only of the immediate present: no past impinges on their actions; no future motivates their thoughts. History, they told Everett, is "where nothing happens, and everything is the same."

Lest anyone should think tribes such as the *Pirahã* somehow lacking in capacity, allow me to mention the *Guugu Yimithirr* of north Queensland in Australia. In common with most Aboriginal language speakers, the *Guugu Yimithirr* have only a handful of number words: *nubuun* (one), *gudhirra* (two), and *guunduu* (three or more). This same language, however, permits its speakers to navigate their landscape geometrically. A wide array of coordinate terms attune their minds intuitively to magnetic north, south, east, and west, so they develop an extraordinary sense of orientation. For instance, a *Guugu Yimithirr* man would not say something like, "There is an ant on your right leg," but rather "There is an ant on your southeast leg." Or, instead of saying, "Move the bowl back a bit," the man would say, "Move the bowl to the north-northwest a bit."

We are tempted to say that a compass, for them, has no point. But at least one other interesting observation can be drawn from the *Guugu Yimithirrs'* ability. In the West, young children often struggle to grasp the concept of a negative number. The difference between the numbers two (2) and minus two (–2) often evades their imagination. Here the *Guugu Yimithirr* child has a definite advantage. For two, the child thinks of "two steps east," while minus two becomes "two steps west." To a question like, "What is minus two plus one?" the Western child might incorrectly offer, "Minus three," whereas the *Guugu Yimithirr* simply takes a mental step eastward to arrive at the right answer of "one step west" (–1).

The *Kpelle* tribe of Liberia offers a final example of culture's effect on how a person counts. The *Kpelle* have no word in their language corresponding to the abstract concept of "number." Counting words exist, but are rarely employed above thirty or forty. One young *Kpelle* man, when interviewed by a linguist, could not recall his language's term for seventy-three. A word meaning "one hundred" frequently stands in for any large amount.

Numbers, the *Kpelle* believe, have power over people and animals and are to be traded lightly and with a kind of reverence; village elders therefore often guard jealously the solutions to sums. From their teachers, the children acquire only the most basic numerical facts in piecemeal fashion, without learning any of the rhythm that constitutes arithmetic. The children learn, for example, that 2 + 2 = 4, and perhaps several weeks or months afterward that 4 + 4 = 8, but they are never required to connect the two sums and see that 2 + 2 + 4 = 8.

Counting people directly is believed by the *Kpelle* to

bring bad luck. In Africa, this taboo is both ancient and widespread. There exists as well the sentiment, shared by the authors of the Old Testament, that the counting of human beings is an exercise in poor taste. The simplicity of their number words is not only a question of language, it also reveals an ethical dimension.

I read with pleasure a book of essays published several years ago by the Nigerian novelist Chinua Achebe. In one, Achebe complained of the Westerners who asked him, "How many children do you have?" Rebuking silence, he suggested, best answered such an impertinent question.

"But things are changing and changing fast with us . . . and so I have learned to answer questions that my father would not have touched with a bargepole."

Achebe's children number *ano* (four). In Iceland, they would say *fjögur.*

Four

PROVERBS AND TIMES TABLES

I once had the pleasure of discovering a book wholly dedicated to the art of proverbs. It was in one of the municipal libraries that I frequented as a teenager. The title of the book escapes me now, the name of its author, too, but I still recall the little shiver of excitement I felt as my fingers caressed its quarto pages.

"Penny wise, pound foolish."
"Small fishes are better than empty dishes."
"A speech without proverb is like a stew without salt."

Now that I come to think of it, I doubt this book had any single author. Every proverb is anonymous. Each appears in a society's mental repertoire by some process akin to immaculate composition. Like the verses in the Muslims' Koran, proverbs seem pre-written, patiently awaiting a mouth to utter their existence. Some linguists contend that language happens independently of our reason, its origins traceable to a still mysterious and exclusive gene. Perhaps proverbial logic is like language in this respect, its existence as essential to our humanity as the power of speech.

Whoever the author or the editor, this book proved to me that there are only so many proverbs a healthy man can take. A point of surfeit is reached, beyond which the reader can no longer follow: his brain starts to ache, his eyes to water. Taken in excess, proverbs lose all the familiar felicities of their compact structure. They start to read merely as repetitions—which many by then quite probably are. From my experience, I estimate this limit at about one hundred.

One hundred proverbs, give or take, sum up the essence of a culture; one hundred multiplication facts compose the ten times table. Like proverbs, these numerical truths or statements—two times two is four, or seven times six equals forty-two—are always short, fixed and pithy. Why then do they not stick in our heads as proverbs do?

But they did before, some people claim. When? In the good old days, of course. Today's children, they suggest, are simply too slack-brained to learn correctly. Nothing interests them but sending one another text messages and harassing the teacher. The critics hark back to those days before computers and calculators; to the time when every number was drummed into children's heads till finding the right answer became second nature.

Except that, no such time has ever really existed. Times tables have always given many schoolchildren trouble, as Charles Dickens knew in the mid-nineteenth century.

> Miss Sturch put her head out of the school-room window: and seeing the two gentlemen approaching, beamed on them with her invariable smile. Then, addressing the vicar, said in her softest tones, "I regret extremely to trouble you, sir, but I find Robert very intractable, this morning, with his multiplication table." "Where does he stick now?" asked Doctor Chennery. "At seven times eight, sir," replied Miss Sturch. "Bob!" shouted the vicar through the window. "Seven times eight?" "Forty-three," answered the whimpering voice of the invisible Bob. "You shall have one more chance before I get my cane," said Doctor Chennery. "Now then, look out. Seven times . . .

Only his younger sister's rapid intervention with the answer—fifty-six—spares the boy the physical pain of another wrong guess.

Centuries old, then, the difficulty that many children face acquiring their multiplication facts is also serious. It is, to borrow a favorite term of politicians, a "real problem." "Lack of fluency with multiplication tables," reports the UK schools inspectorate, "is a significant impediment to fluency with multiplication and division. Many low-attaining secondary-school pupils struggle with instant recall of tables. Teachers [consider] fluent recall of multiplication tables as an essential prerequisite to success in multiplication."

The facts in a multiplication table represent the essence of our knowledge of numbers: the molecules of math. They tell us how many dimes make up a dollar (10×10), the number of squares on a chessboard (8×8), the quantity of individual surfaces on a trio of boxes (3×6). They help us evenly divide fifty-six items among eight people ($7 \times 8 = 56$, therefore $56/8 = 7$), or realize that forty-three of something cannot be evenly distributed in the same way (because forty-three, being a prime number, makes no appearance among the facts). They reduce the risk of anxiety in the young learner, and give a vital boost to the child's confidence.

Patterns are the matter that these molecules, in combination, make. Take, for instance, the consecutive facts $9 \times 5 = 45$, and $9 \times 6 = 54$: the digits in both answers are the same, only reversed. Thinking about the other facts in the nine times table, we see that every answer's digits sum to nine:

$9 \times 2 = 18$ ($1 + 8 = 9$)

$9 \times 3 = 27$ ($2 + 7 = 9$)

$9 \times 4 = 36$ ($3 + 6 = 9$)

Etc. . . .

Or, surveying the other tables, we discover that multiplying an even number by five will always produce an answer ending in zero ($2 \times 5 = 10$, $6 \times 5 = 30$), while multiplying an odd number by five gives answers that always end in itself ($3 \times 5 = 15$, $9 \times 5 = 45$). Or, we spot that six squared (thirty-six) plus eight squared (sixty-four) equals ten squared (one hundred).

Sevens, the trickiest times table to learn, also offer a beautiful pattern. Picture the seven on a telephone's keypad, in the bottom left-hand corner. Now simply raise your eye to the key immediately above it (four), and then again to the next key above (one). Do the same starting from the bottom middle key (eight), and so on. Every keypad digit in turn corresponds to the final digit in the answers along the seven times table: 7, 14, 21, 28...

Not all multiplication facts pose problems, of course. Multiplying any number by one or ten is obviously easy enough. Our hands know that two times five, and five times two, both equal ten. Equivalencies abound: two times six, and three times four, both lead to twelve; multiplying three by ten, and six by five, amounts to the same thing.

But others are trickier, less intuitive, and far easier to let slip. A numerate culture will find whatever means at its disposal to pass these obstinate facts down from one generation to the next. It will carve them into rock and scratch them onto parchment. It will condemn every inauspicious student to threats and thrashings. It will select the most succinct form and phrasing for its essential truths: not too heavy for the tongue, nor too lengthy for the ear.

Just like a proverb.

For example, what did our ancestors mean precisely when bequeathing us a truth like "An apple a day keeps the doctor away"? Not, of course, that we should read it literally, superstitiously, imagining apples like the cloves of garlic that are supposed to make vampires take to their heels. Rather, the sentence expresses a core relationship between two different things: healthy food (for which the apple plays stand-in), and illness (embodied by the doctor). Consider a few of the alternate ways in which this relationship might also have been summed up:

"A daily fruit serving is good for you."

"Eating healthy food prevents illness."

"To avoid getting sick, eat a balanced diet."

These versions are as short, or even shorter, than our proverb. But none is anywhere near as memorable.

Long before Dickens wrote about the horrors of multiplication tables, our ancestors had decided to sum up fifty-six as "seven times eight," just as they described health (and its absence) in apples and doctors. But as with a concept like "health," understanding the number fifty-six can be achieved via many other routes.

$56 = 28 \times 2$

$56 = 14 \times 4$

$56 = 7 \times 8$

Or even:

$56 = 3.5 \times 16$

$56 = 1.75 \times 32$

$56 = 0.875 \times 64$

It is not difficult to see, though, why tradition would have privileged the succinctness and simplicity of "seven times eight" for most purposes, over rival definitions such as "one and three-quarters times thirty-two" or "seven-eighths of sixty-four" (as useful as they might be in certain contexts).

What is seven times eight? It is the clearest and simplest way to talk about the number fifty-six.

These familiar forms may be simple and succinct, but they are finely wrought, nonetheless, whether in words or figures. The proverbial apple, for example, begins the proverb, though its meaning (as a protector of health) cannot be grasped until the end. "Apple" here is the answer to the question: What keeps the doctor away? Other proverbs also share this structure, where the answer precedes the question. "A stitch in time saves nine" (What saves nine stitches? A stitch in time) or "Blind is the bookless man" (What is the bookless man? Blind).

Placing the answer at the start compels our imagination: we concede more freely the premise that an apple can deter illness, in part because the word "apple" precedes all the others. Using this structure can also arouse our attention, inciting us to picture the rest of the proverb with the opening image in mind: to see the bookless man, for example, more clearly in light of his blind eyes.

When I discussed the ways in which we could think about the number fifty-six, I borrowed this feature of proverbs and put the sum's answer at the start. Saying, "Fifty-six equals seven times eight" lends emphasis where it is needed most: not on the seven or the eight, but on what they produce.

Form is important. A pupil reads $56 = 7 \times 8$ and hears the whisper of many generations, while another child, shown $7 \times 8 = 56$, finds himself alone. The first child is enriched; the second is disinherited.

Today's debates over times tables often neglect questions of form. Not so the schools of nineteenth-century America. The young nation, still younger than its oldest citizens, hosted educational discussions unprecedented in their inquisitive detail. Teachers pondered in marvelous depth the kind of verb to use when multiplying. In *The Grammar of English Grammars* (published in 1858) we read, "In multiplying one only, it is evidently best to use a singular verb: 'Three times one is three.' And in multiplying any number above one, I judge a plural verb to be necessary: 'Three times two are six.'"

The more radical contributors to these debates suggested doing away with excess words like "times" altogether. Instead of learning "four times six is twenty-four," the child would repeat, "four sixes are twenty-four." These educators urged a return to the way ancient Greek children had chanted their times tables two millennia before: "once one is one," "twice one is two," et cetera. Others went even further, suggesting that the verb "is" (or "are") also be thrown out: "four sixes, twenty-four," in the manner of the Japanese.

Schools in Japan have long lavished attention on the sounds and rhythms of the times tables. Every syllable counts. Take the multiplication $1 \times 6 = 6$, among the first facts that any child learns. The standard Japanese word for "one" is *ichi;* the usual Japanese word for "six" is *roku.* Put together, they make: *ichi roku roku* (one six, six). But Japanese pupils never say this: the line is clumsy, the sounds cacophonous. Instead, the pupils all say, *in roku ga roku* (one six, six), using an abraded form of *ichi* (*in*) and an insertion (*ga*) for euphony.

The trimming of unnecessary words or sounds shapes both the proverb and the times table. "Better late than never," says the parent in New York when his son complains of his allowance long overdue. "Four fives, twenty," says the boy when he finally counts his money.

In Japanese, the multiplication $6 \times 9 = 54$ is an extreme example of ellipsis. Being similar in sound, the two words—*roku* (six) and *ku* (nine)—merge to a single *rokku.* This new number would be a little like pronouncing the multiplication 7×9 in English as "sevine."

Why is *in roku ga roku* judged more pleasing than *ichi roku roku?* Both phrases contain six syllables; both phrases use the "roku" word twice, yet the first sounds beautiful, while the second seems ugly. The answer is, parallelism. *In roku ga roku* has a parallel structure, which makes it easier on the ear. We hear this balanced structure frequently in our proverbs: "fight fire with fire." One six is six.

It is much harder to make good parallel times tables in English than it is in Japanese. The same is true of many

European languages. In Japanese, a child says *roku ni juuni* (six two, ten two) for 6 × 2 = 12, and *san go juugo* (three five, ten five) for 3 × 5 = 15, whereas an English-speaking child must say "twelve" and "fifteen," a French child *douze* and *quinze,* and a German child *zwölf* and *fünfzehn*.

Beside the ones, it is only the ten times table that produces consistent parallel forms in English, of the kind "easy come, easy go": "seven tens (are) seventy."

Not all proverbs employ parallels. Many use alliteration—the repetition of certain sounds: "One swallow does not a summer make" or "All that glitters is not gold." Times tables alliterate too: "four fives (are) twenty" and (if we extend our times tables to twelve) "six twelves (are) seventy-two."

Parallels and alliteration are both in evidence when proverbs rhyme: "A friend in need is a friend indeed" or "Some are wise and some are otherwise." By definition, square multiplications (when the number is multiplied by itself) begin in a similar way: "two times two..." "four fours..." "nine times nine..." though only the squares of five and six finish with a flourish: "five times five (is) twenty-five" and "six times six (is) thirty-six."

For this reason, learners procure these two multiplication facts (after, perhaps, "two times two is four") with the greatest ease and pleasure. This pair—"five times five is twenty-five" and "six times six is thirty-six"—truly attain the special quality of the proverb. Other multiplication facts, from the same times tables, get close. For example, multiplying five by any odd number inevitably leads to a rhyme: "seven fives (are) thirty-five." Six, when followed by an even number, causes the even number to rhyme: "six times four (is) twenty-four" and "six eights (are) forty-eight."

Mistakes? Of course they happen. Nobody is above them. No matter how long a person spends immersed in numbers, recollection can sometimes go astray. I have read of world-class mathematicians who blush at "nine times seven."

The same troubles we encounter with times tables sometimes occur with words, what we call "slips of the tongue," though more often than not the tongue is innocent. It is the memory that is to blame. Someone who says, "He is like a bear with a sore thumb" (mixing up "like a bear with a sore head" and "to stick out like a sore thumb") makes a mistake similar to he who answers "seven times eight" with "forty-eight" (confusing $7 \times 8 = 56$ with $6 \times 8 = 48$).

Such mistakes are mistakes of unfamiliarity. Proverbs, like times tables, can often strike us as strange, their meanings remote. Why do we talk of bears with sore heads? In what way do swallows conjure up summertime better than other birds? The choice of words seems to us as arbitrary and archaic as the numbers in the times tables. But the truths they represent are immemorial.

"Hold fast to the words of ancestors," instructs a proverb from India. Hold fast to their times tables, too.

Five

CLASSROOM INTUITIONS

Television journalists, in their weaker moments, will occasionally pull the following stunt on a hapless minister of education. Mugging through his makeup at the attendant cameras, the interviewer strokes his notes, clears his throat, and says, "One final question. What is nine times six?"

Such episodes never fail to make me sigh. It is a sad thing when mathematics is reduced to the recollection (or, more often, the non-recollection) of a classroom rule.

In one particular confrontation of this type, the presenter demanded to know the price of fourteen pens when four had a price tag of $2.42. "I haven't the foggiest," the minister whimpered to the audience members' howls of delight.

Of course, the questions are patently asked with the expectation of failure in mind. Politicians are always trying to anticipate our expectations and to meet them. Should we then feel such surprise when they judge the situation correctly and get the sum wrong?

Properly understood, the study of mathematics has no end: the things each of us does not know about it are infinite. We are all of us at sea with some aspect or another. Personally, I must admit to having no affinity with algebra. This discovery I owe to my middle school math teacher, Mr. Baxter.

Twice a week, I would sit in Mr. Baxter's class and do my best to keep my head down. I was thirteen, going on fourteen. With his predecessors I had excelled at the subject: number theory, statistics, probability, none of them had given me any trouble. Now I found myself an algebraic zero.

Things were changing; I was changing. All swelling limbs and sweating brain, suddenly I had more body than I knew what to do with. Arms and legs became the prey of low desktops and narrow corridors, were ambushed by sharp corners. Mr. Baxter ignored my plight. Bodies were inimical to mathematics, or so we were led to believe. Bad hair, acrid breath, lumpy skin, all vanished for an hour every Tuesday and Thursday. Young minds in the buff soared into the sphere of pure reason. Pages turned into parallelograms; cities, circumferences; recipes, ratios. Shorn of our bearings, we groped our way around in this rarefied air.

It was in this atmosphere that I learned the rudiments of algebra. The word, we were told, was of Arabic origin, culled from the title of a ninth-century treatise by Al-Khwarizmi ("algorithm," incidentally, is a Latin corruption of his name). This exotic provenance, I remember, left a deep impression on me. The snaking, swirling equations in my textbook made me think of calligraphy. But I did not find them beautiful.

My textbook pages looked cluttered with lexicographical debris: all those x's and y's and z's. The use of the least familiar letters only served to confirm my prejudice. I thought these letters ugly, interrupting perfectly good sums.

Take $x^2 + 10x = 39$, for example. Such concoctions made me wince. I much preferred to word it out: a square number (1, or 4, or 9, etc.) plus a multiple of ten (10, 20, 30, etc.) equals thirty-nine; 9 (3 × 3) + 30 (3 × 10) = 39; three is the common factor; $x = 3$. Years later, I learned that Al-Khwarizmi had written out all his problems, too.

Stout and always short of breath, Mr. Baxter had us stick to the exercises in the book. He had no patience whatsoever for paraphrasing. Raised hands were cropped with a frown and the admonition to "reread the section." He was a stickler for the textbook's methods. When I showed him my work, he complained that I had not used them. I had not subtracted the same values from either side of the equation. I had not done a thing with the brackets. His red pen flared over the carefully written words of my solutions.

Let me give a further example of my deviant reasoning: $x^2 = 2x + 15$. I word it out like this: a square number (1, 4, 9, etc.) equals fifteen more than a multiple of two (2, 4, 6, etc.). In other words, we are looking for a square number above seventeen (being fifteen more than two). The first candidate is twenty-five (5 × 5) and twenty-five is indeed fifteen more than 10 (a multiple of two); $x = 5$.

A few of Mr. Baxter's students acquired his methods; most, like me, never did. Of course I cannot speak for the others, but I found the experience bruising. I was glad when the year was over and I could move on to other math. But I also felt a certain shame at my failure to comprehend. His classes left me with a permanent suspicion of all equations. Algebra and I have never been fully reconciled.

From Mr. Baxter I learned at least one profitable lesson: how not to teach. This lesson would serve me well on numerous occasions. Two years after I left school, while leafing through the newspaper one morning, I came across an agency's advertisement recruiting tutors. I had taught English in Lithuania and discovered that I liked teaching. So I applied. I was interviewed by a lady up in years named Grace, in the office that she kept in her living room. Needlepoint cushions filled the small of my back as I sat before her desk. The wallpaper, if I remember correctly, had a pattern of small birds and honeybees.

The meeting was brief.

"Do you enjoy helping others to learn something new?"

"Are you mindful of a student's personal learning style?"

"Could you work according to a set curriculum?"

Her questions contained their own answers, like the dialogue taught in a foreign language class: "Yes, I could work according to a set curriculum."

After ten minutes of this she said, "Excellent, well, you would certainly fit in with us here. For English we already have tutors and there is not much demand for foreign languages. What about primary school–level math?"

What about it? I was a taker.

Grace's bookings certainly kept me on my toes. My tutoring patch extended to the neighboring town five miles away, and the bus ride and walk to the farthest homes took as long as the lessons. Nervous, I learned on the job, but the families helped me. I found the children, between the ages of seven and eleven, generally polite and industrious; their parents' nods and smiles tamed my nerves. After a while I ceased to worry and even began to look forward to my weekly visits.

Should I admit I had a favorite student? He was a brown-haired, freckled boy, eight years old but small for his age, and the first time I came to the house he fairly shivered with shyness. We started out with the textbooks that the agency loaned me, but they were old and smelly and the leprous covers soon came apart in our hands. A brightly colored book replaced them, one of the boy's Christmas gifts, but its jargon was poison to his mind. So we abandoned the books and found some better way to pass the hour together. We talked a lot.

It turned out that he had a fondness for collecting baseball cards and could recite the names of the players depicted on them by heart. With pride he showed me the accompanying album.

"Can you tell me how many cards you have in there?" I asked him. He admitted to having never totaled them up. The album contained many pages.

"If we count each card one by one it will take quite a long time to reach the last page," I said. "What if we were to count them two by two instead?" The boy agreed that would be quicker. Twice as quick, I pointed out, "And what if we were to count the cards in threes? Would we get to the end of the album even faster?" He nodded. Yes, we would: three times as fast.

The boy interrupted. "If we counted the cards five at a

time, we would finish five times faster." He smiled at my smile. Then we opened the album and counted the cards on the first page, I placing my larger palm over every five. There were three palms of cards: fifteen. The second page had slightly fewer (two palms and three fingers, thirteen)—so we carried the difference over to the next. By the seventh page we had reached twenty palms: one hundred cards. We continued turning the pages, putting my palm onto each, and counting along. In all, the number of cards rose to over eighty palms (four hundred).

After making light work of the album, we considered the case of a giant counting its pages. The giant's palm, we agreed, would easily count the cards by twenty.

What if the same giant wanted to count up to a million? The boy thought for a moment. "Perhaps he counts in hundreds: one hundred, two hundred, three hundred..." Did the boy know how many hundreds it would take to reach a million? He shook his head. Ten thousand, I told him. His eyebrows leaped. Finally, he said, "He would count in ten thousands, then, wouldn't he?" I confirmed that he would: it would be like us counting from one to hundred. "And if he were really big, he might count in hundred thousands," I continued. The giant would reach a million as quickly as we counted from one to ten.

Once, during a lesson solving additions, the boy hit upon a small but clever insight. He was copying down his homework for us to go through together. The sum was 12 + 9, but it came out as 19 + 2. The answer, he realized, did not change. Twelve plus nine, and nineteen plus two, both equal twenty-one. The fortuitous error pleased him; it made him pause and think. I paused as well, not wishing to talk for fear of treading on his thoughts. Later I asked him for the answer to a much larger sum, something like 83 + 8. He closed his eyes and said, "Eighty-nine, ninety, ninety-one." I knew then that he had understood.

Of the other students, I recall the Singh family whose children I taught on Wednesday evenings for two hours back-to-back. I remember I could never get on with the father, a businessman who put on executive airs, although the mother treated me with nothing but kindness. They had three children, two boys and a girl; the children were always waiting for me around the table in the living room, still dressed up in their school uniforms. The eldest boy was eleven: a bit of a show-off, he had the confidence of an eldest son. His sister regularly deferred to him. In the middle, the second son laughed a lot. He seemed to laugh for the whole family.

In the beginning, the trio took the wan bespectacled man before them only half seriously as a teacher. Between them, they had ten years on me. I looked too young and probably sounded it as well, without the smooth patter that comes with experience. All the same, I held my ground. I helped them with their times tables, in which they were far from fluent. They showed surprise when I failed to berate their errors and hesitations. On the contrary, if they were close to the right answer I told them so.

"What is eight times seven?"

"Fifty..." The eldest boy wavered.

"Yes," I said, encouraging.

"Fifty-*four,*" he ventured.

"Nearly," I said. "Fifty-six."

This hesitation, a habit of many of my students, intrigued me. It suggested not ignorance, but rather indecision. To say that a student has no idea of a solution, I realized, is untrue. Truth is, the learner *does* have ideas, too many, in fact—almost all of them bad. Without the knowledge necessary to eliminate this mental haze, the learner finds himself confronted with an embarrassment of wrong answers to helplessly pick from.

What had the boy been thinking, I inquired, when he selected fifty-four as his answer? He admitted having previously considered fifty-three, fifty-six, fifty-seven, and fifty-five (in that order). He had felt sure that fifty-one or fifty-two would be too small an answer, both fifty-eight and fifty-nine, too high. Then I asked him why he had finally preferred fifty-four to fifty-three. He replied that he had thought of the eight in the question, and the fact that fifty is half of one hundred, and that half of eight is four.

We moved on to discussing the difference between odd and even numbers. Eight was an even number; seven was odd. What happens when we multiply odd by even? The children's faces showed more hesitation. I suggested several examples: two times seven (fourteen), three times six (eighteen), four times five (twenty), every answer an even number. Could they see why? Yes, said the second son at last: multiplying by an even number was the same thing as creating pairs. Two times seven created a pair of sevens; four times five made two pairs of fives; three times six produced three pairs of threes. What, then, about eight times seven? It meant four pairs of sevens, the boy said.

Pairs evened every odd number: one sock became two socks, three became six, five became ten, seven became fourteen, and nine became eighteen. The last digit of a pair would always be an even number.

The mental fog of eight times seven was evaporating. Fifty-three promptly vanished as a contending answer; so too did fifty-seven and fifty-five. That left fifty-four and fifty-six. How, then, to tell the two apart? The number fifty-four, I pointed out, was six away from sixty: fifty-four—like sixty—was divisible by six. Like sixty, fifty-four will therefore be the answer to a question containing a six (or a number divisible by six), but neither a seven nor an eight.

By this process of elimination, which amounts to careful reasoning, only fifty-six remained. Fifty-six is two sevens away from seventy, three eights away from eighty. Eight times seven equals fifty-six.

My only adult student was a housewife with copper-colored skin, and a long name with vowels and consonants that I had never before seen in such a permutation. The housewife, Grace informed me on the phone, aspired to professional accountancy. In my mind, this fact made for an unpromising start. The admission of self-interest rubbed up against my naive vision of mathematics as something playful and inventive. There seemed to me something almost vulgar in the housewife's sudden interest in numbers, as if she wanted to befriend them only as some people set out to befriend well-connected people.

Quickly I amended this faulty judgment. The reservations about my new student were unfair. They were the reservations of a children's tutor—I knew nothing of teaching adults, of anticipating their needs and expectations.

We were talking together one day about negative numbers, as we sat in her white-tiled kitchen. Like mathematicians of the sixteenth century, who referred to them as "absurd" and "fictitious," she found these numbers difficult to imagine. What could it mean to subtract something from nothing? I tried to explain, but found myself at a frustrating loss for words. Somehow my student understood.

"You mean, like a mortgage?"

I did not know what a mortgage was. Now it was her turn to try to explain. As she spoke, I realized that she knew a lot more about negative numbers than I did. Her words possessed real value: they had the bullion of hard experience to back them up.

Another time we went over "improper" fractions: top-heavy fractions, like four-thirds (4/3) or seven-quarters (7/4), that help us to think about units in different ways. If we think of the number one as being equivalent to three thirds, for example, then four-thirds is another way of describing one and one-third. Seven-quarters, my student and I agreed, resembled two apples that had been quartered (considering "one" as equal to four quarters) and one of these eight apple-quarters consumed.

Our hour was soon up, but we kept on talking. We were discussing fractions and what happens when you halve a half of a half of a half, and so on. It amazed us both to think that this halving, theoretically speaking, could continue indefinitely. There was pleasure in confiding our mutual amazement, almost in the manner of gossip. And like gossip, it was something that we both knew and did not know.

Then she came to a beautiful conclusion about fractions that I shall never forget.

She said, "There is no thing that half of it is nothing."

Six

SHAKESPEARE'S ZERO

Few things, to judge by his works, so fascinated William Shakespeare as the presence of absence: the lacuna where there ought to be abundance—of will, or judgment, or understanding. It looms large in the lives of many of his characters, so powerful in part because it is universal. Not even kings are exempt.

> LEAR: What can you say to draw
> A third more opulent than your sisters? Speak.
> CORDELIA: Nothing, my lord.
> LEAR: Nothing?
> CORDELIA: Nothing.
> LEAR: Nothing will come of nothing. Speak again.

The scene is one of the tensest, most suspenseful moments in theater, a concentration of tremendous force within a single word. It is the ultimate negation, tossed between the old king and his beloved youngest daughter, compounded and multiplied through repetition. Nothing. Zero.

Of course, Shakespeare's contemporaries were familiar

with the idea of nothingness, but not with nothingness as a number, something that they could count and manipulate. In his arithmetic lessons, William became one of the first generation of English schoolboys to learn about the figure zero. It is interesting to wonder about the consequences of this early encounter. How might the new and paradoxical number have driven his thoughts along particular paths?

Arithmetic spelled trouble for many schoolmasters of the period. Their knowledge of it was often suspect. For this reason, lessons were probably short, often kept until almost the final hour of the afternoon. Squeezed in after long bouts of Latin composition, a list of proverbs, or the chanting of prayers, the sums and exercises were mostly drawn from a single textbook: *The Ground of Artes* by Robert Recorde. Published in 1543 (and again, in an expanded edition, in 1550), Recorde's book, which included the first material on algebra in the English language, taught "the work and practice of Arithmetike, both in whole numbres and Fractions after a more easyer and exarter sorte than any like hath hitherto been sette furth."

Shakespeare learned to count and reckon using Recorde's methods. He learned that "there are but tenne figures that are used in Arithmetick; and of those tenne, one doth signifie nothing, which is made like an O, and is privately called a Cypher." These Arabic numbers—and the decimal place system—would soon eclipse the Roman numerals (the Tudors called them "German numbers"), which were often found too cumbersome for calculating.

German numbers were, of course, letters: I (or j), one; V,

five; X, ten; L, fifty; C, one hundred; D, five hundred; and M, one thousand. Six hundred appeared as vj.C and three thousand as CCC.M. Perhaps this is why Recorde compares the zero to an O. Years later, Shakespeare would deploy the cipher to blistering effect. "Thou art an O without a figure... Thou art nothing," the Fool tells Lear, after the dialogue with Cordelia that destroys the king's peace of mind.

In Shakespeare's lessons, letters were out, figures (digits) in. Perhaps they were displayed conspicuously on charts, hung up on walls like the letters of the alphabet are now. Ten to a hard bench, the boys trimmed and dipped their quills in ink, and their feathered fingers copied out the numbers in small, tidy lines. Zeroes spotted every page. But why record something that has no value? Something that is nothing?

England's medieval monks, translators and copiers of earlier treatises by the Arab world's mathematicians, had long before noted the zero's almost magical quality. One twelfth-century scribe suggested giving it the name "chimera" after the fabulous monster of Greek mythology. Writing in the thirteenth century, John of Halifax explained the zero as something that "signifies nothing" but instead "holds a place and signifies for others." His manuscript proved popular in the universities. But it would take the invention of Gutenberg's printing press to bring his ideas to a much wider audience, including the motley boys of King's New School in Stratford.

GLOUCESTER: What paper were you reading?
EDMUND: Nothing my lord.

> GLOUCESTER: No? What needed then that terrible despatch of it into your pocket? The quality of nothing hath not such need to hide itself. Let's see: Come, if it be nothing I shall not need spectacles.

The quality of nothing. We can picture the proto-playwright grappling with zero. The boy closes his eyes and tries to see it. But it is not easy to see nothing. Two shoes, yes, he can see them, and five fingers and nine books. 2 and 5 and 9: he understands what they mean. How, though, to see *zero* shoes? Add a number to another number, like a letter to another letter, and you create something new: a new number, a new sound. Only, if you add a number to zero, nothing changes. The other number persists. Add five zeroes, ten zeroes, a hundred zeroes if you like. It makes no difference. A multiplication by zero is just as mysterious. Multiply a number, any number—three, or four hundred, or 5,678—by zero, by nothing, and the answer is zero.

Was the boy able to keep up in his lessons, or did he lag? The Tudor schoolmaster, violence dressed in a long cloak and black shoes, would probably have concentrated his mind. The schoolmaster's cane could reduce a boy's buttocks to one big bruise. With its rhyming dialogues (even employing the occasional joke or pun) and clear examples intended to give "ease to the unlearned," we can only hope that Recorde's book spared Shakespeare and his classmates much pain.

> Heere you see vj. (six) lines, whiche stande for vj. (six) places... the [lowest line] standeth for the first place,

> and the next above it for the second, and so upwarde, till you come to the highest, whiche is the sixth line, and standeth for the sixth place.
>
> 100000
>
> 10000
>
> 1000
>
> 100
>
> 10
>
> 1
>
> The first place is the place of units or ones, and every counter set in that line betokeneth but one, and the seconde line is the place of 10, for everye counter there standeth for 10. The thirde line the place of hundreds, the fourth of thousands, and so forth.

Perhaps talk of counters turned the boy's thoughts to his father's glove shop. His father would have accounted for all his transactions using the tokens. They were hard and round and very thin, made of copper or brass. There were counters for one pair of gloves, and for two pairs, and three and four and five. But there was no counter for zero. No counters existed for all the sales that his father did not close.

I imagine the schoolmaster sometimes threw out questions to the class. How do we write three thousand in digits? From Recorde's book, Shakespeare learned that zero denotes size. To write thousands requires four places. We write: 3 (three thousands), 0 (zero hundreds), 0 (zero tens), 0 (zero units): 3,000. In *Cymbeline,* we find one of Shakespeare's many later references to place value (what Recorde in his book called each digit's "roome").

> Three thousand confident, in act as many—
> For three performers are the file, when all
> The rest do nothing—with this word "Stand, stand,"
> Accommodated by the place...

The concept must have fascinated him ever since his schooldays. A nothing, the boy sees, depends on kind. An empty hand, for example, is a smaller nothing than an empty class or shop, in the same way that the zero in 10 is ten times smaller than the zero in 101. And the bigger the number, the more places and therefore the more zeroes it can contain: ten has one zero, whereas one hundred thousand has five. He thinks, the bigger an empty room, the more the things that can be contained inside: the greater an absence, the greater the potential presence. Subtract (or "rebate" as the Tudors said) one from one hundred thousand and the entire number transforms: five zeroes, five nothings, all suddenly turn to nines (the largest digit): 99,999. Perhaps, like Polixenes in *The Winter's Tale,* he sensed already the tremendous potential of self-effacement, understood imagination as shifting from place to place like a zero inside an immense number.

And therefore, like a cipher [zero]
(Yet standing in rich place), I multiply
With one "We thank you" many thousands more
That go before it

Recorde's book abounded with exercises. Shakespeare and his classmates' sheets of paper would have quickly turned black with sums as they measured cloth and purchased loaves and counted sheep and paid clergy. But William's mind returns ceaselessly to the zero. He thinks of ten and how it differs from his father's ten. To his father, ten (X) is twice five (v): he counts, whenever possible, in fives and tens. To his son, ten (10) is a one (1) displaced, accompanied by a nought. To his father, ten (X) and one (i) have hardly anything in common: they are two values on opposite ends of a scale. But for the boy, ten and one are intimately linked: there is nothing between them.

Ten and one, one and ten.

The boy sees that, with a retinue of zeroes, even the humble one becomes enormously valuable. Imagination can reconcile even one and one million, as Shakespeare affirmed in his prologue to *Henry V*, when the chorus stake their claim to represent the multitudes on the field of Agincourt.

O, pardon! Since a crooked figure [digit] may
Attest in little place a million,
And let us, ciphers to this great accompt,
On your imaginary forces work

But it is perhaps in his poetry that the boy—now a young man—would most clearly express the impact of Recorde's teaching on his mind. In Sonnet 38, Shakespeare wrote about the relationship between himself and his beloved Muse, comparing their couple to a ten: the poet, the zero, and his beloved, the one.

> O, give thyself the thanks, if aught in me
> Worthy perusal stand against thy sight...
> Be thou the tenth Muse, ten times more in worth...

This relationship, as everyone knows, would prove remarkably fruitful: his poems and plays multiplied. In the Globe Theatre, round as an O, an empty cipher filled with meaning, Shakespeare's loquacious quill drew crowds.

"Shakespeare was the least of an egotist that it was possible to be," wrote the nineteenth-century critic William Hazlitt. "He was nothing in himself; but he was all that others were, or that they could become." That nothing, who had once been a dazzled schoolboy, laboring for the breakthrough moment when he comprehended the paradoxical fullness of the empty figure of a zero, would surely have been delighted with this description.

Seven

SHAPES OF SPEECH

We know next to nothing with any certainty about Pythagoras, except that he was not really called Pythagoras. The name by which he is known to us was probably a nickname bestowed by his followers. According to one source, it meant "He who spoke truth like an oracle." Rather than entrust his mathematical and philosophical ideas to paper, Pythagoras is said to have expounded them before large crowds, making the world's most famous mathematician its first rhetorician.

It is easy to imagine the atmosphere of intrigue and anticipation that would have attended such a spectacle. To believe later accounts, people came from far and wide to hear the legendary figure speak. Citizen after citizen: male and female, young and old, rich and poor, politicians and lawyers and doctors and housewives and poets and farmers and children. Latecomers, red-faced from running, jostled for a place at the back. Waiting for the event to start, they would have shared gossip. Pythagoras, someone whispered, has a thigh made of gold. His words can soothe even wild bears, said another. He communes with nature, affirmed a third. Even the rivers know his name!

Pythagoras was a handsome, forty-something-year-old man when, in around 530 BCE, he founded his school of disciples on the Greek colony of Croton in southern Italy. The residents of this distant outpost, many hundreds of miles from Athens, treated the newcomer's teachings with the highest respect. Their appetite for novelty, excitement, "the next big idea," would have been considerable. Prestige, and perhaps some educational and economic advantage over the neighboring colonies, was also on many minds.

By all accounts, Pythagoras's ideas were more than up to his students' expectations. Mathematics, for him, was nothing less than a way of life. He "transformed the study of geometry into a liberal education," wrote Proclus, the last major Greek philosopher, "examining the principles of the science from the beginning and probing the theorems in an immaterial and intellectual manner." Pythagoras is reputed to have taught that the identities of all existing objects depended on their form rather than on their substance, and could consequently be described using numbers and ratios. The entire cosmos constituted some vast and glorious musical scale. The Pythagoreans thus became the first to understand the world not via tradition (religion), or observation (empirical data), but through imagination—the prizing of pattern over matter.

Pythagoras had star quality, that much is clear. His timing was impeccable. An unhurried speaker, he took his time before addressing the eager crowd. Everyone had the feeling that Pythagoras was speaking to him alone. Not a single phrase went above the listener's head. He understood completely. "Yes," he would think to himself, "yes, it is just as he says. It cannot be any other way." But of course this flash of certainty, this grasping of an absolute truth, was an illusion. The listener's mind had duly accompanied the speaker down one carefully laid path of reasoning, though many alternate paths existed. The listener simply forgot about those other ways of thinking and seeing the world. The audience was swept, logical step by logical step, from their old certainties to new and unexpected ones. This was the power of Pythagoras.

Rhetoric, the art of speaking, gave shape and solidity to Pythagoras's words and ideas. It would also mark the beginning of truly mathematical thought. "A [mathematical] proof," says Steven G. Krantz, a mathematician based at Washington University in St. Louis, "is a rhetorical device for convincing someone else that a mathematical statement is true or valid." Philip J. Davis and Reuben Hersh, two other leading academics, support this view. "Mathematics in real life is a form of social interaction where 'proof' is a complex of the formal and the informal, of calculations and casual comments, of convincing argument and appeals to the imagination and the intuition."

Ancient Greece, with its passionate debates, litigious citizens, and rowdy assemblies, proved an ideal environment for just this kind of social interaction, and consequently, for the development of both rhetoric and mathematics. In fact, without the refinement of rhetoric, there would have been no logic, and thence none of the mathematics that forms one of the cornerstones of our empirically minded Western civilization. Before any of these cultural and intellectual endeavors, there was the practice of persuasion, through argument and the evaluation of evidence. And it was in the law courts, with their public trials, that these building blocks of our system of thought were honed.

Court cases were a daily occurrence in ancient Athens. Hundreds, sometimes thousands, of free citizens filled theaters to hear the parties speak. These citizens composed a vast and anonymous jury: only their age (above thirty years old) and maleness united them. Since the jury was always made up of an odd number of people, no tied judgments were possible. Every decision was final, and without appeal.

In the space of several hours, the defendant and his accuser took center stage. The accuser would speak first, leaving the defendant to refute his arguments. Each wanted to be a Pythagoras. Each tried to dazzle the men of the jury with the order, rhythm, and precision of his words. For the less gifted or confident, eloquence could be bought at a price; professional speechwriters were very much in demand. A single tightly argued presentation, every Greek knew, could make the difference between liberty and imprisonment, between life and death.

Let us picture a trial. The defendant, it is claimed, killed the accuser's son for his one hundred gold coins. What argument does the grieving father bring to the court? Perhaps he points to a similar case, known to everyone present, of a man who killed another for his ten gold coins. If a man would be prepared to risk his skin for ten coins, says the father, then he would surely risk it for one hundred coins. The father's argument, by which he establishes motive, is an exercise in basic mathematical logic: if x is true, then x^2 is also true.

A real example of ancient Greek court rhetoric has come down to us from the orator Antiphon's *First Tetralogy.* It involves the case of a man accused of killing his victim (along with the victim's slave) in cold blood. Anticipating a likely defense of "someone else did it," the plaintiff methodically goes through the possible scenarios, eliminating each in turn, in a style of argument that resembles what mathematicians call "proof by exhaustion."

> Malefactors [thieves] are not likely to have murdered him, as nobody who was exposing his life to a very grave risk would forgo the prize when it was securely within his grasp; and the victims were found still wearing their cloaks. Nor again did anyone in liquor kill him; the murderer's identity would be known to his boon companions. Nor again was his death the result of a quarrel; they would not have been quarreling at the dead of night or in a deserted spot. Nor did the criminal strike the dead man when intending to strike someone else; he would not in that case have killed master and

slave together. As all grounds for suspecting that the crime was unpremeditated are removed, it is clear from the circumstances of death themselves that the victim was deliberately murdered.

The plaintiff renders the potential defense precisely as: "death either by (1) thieves (2) drunks (3) quarrel (4) accident" in order to refute each part. But he does something more. Each scenario represents a miniature proof. Rejecting, for instance, the possibility that a thief committed the murder, his argument proceeds:

A thief (claims the defendant) killed the victim.

But thieves would steal the victim's cloak.

Thus a thief did not kill the victim.

Which self same structure we can find throughout the *Elements* of Euclid.

CA and CB are each equal to AB.

But things equal to the same thing are also equal to one another.

Thus CA is also equal to CB.

Written in the third century before the Christian era, it is hard to overstate the *Elements'* importance to the history of intellectual progress, or the extent to which it embodied the flourishing of rhetoric and logic that permitted its creation. Proposition 21, in Book IX of the treatise (among its oldest pages, dating back to the Pythagoreans), illustrates the lawyerly manner that its author adopted throughout.

> If as many even numbers as we please be added together, the whole is even . . . For, since each of the numbers . . . is even, it has a half part; so that the whole [number] also has a half part. But an even number is that which is divisible into two equal parts; therefore [the whole number] is even.

We might summarize this argument as:

> Proposition: Even numbers (of any quantity) added together make an even number.
>
> Clarification: Since even numbers have half parts, their sum will also have a half part.
>
> Axiom: An even number is that which is divisible into two equal parts.
>
> Conclusion: Therefore the sum of any quantity of even numbers is even.
>
> This echoes any number of the possible arguments that would have passed through the Ancient Greek courts.
>
> Proposition: The accused stole my ox.

> Clarification: Since he said nothing to me about the ox before taking it, the ox was taken without my permission.
>
> Axiom: An item of property removed without the owner's consent is stealing.
>
> Conclusion: Therefore my ox was stolen.

With their axioms—statements that we accept as being self-evidently true—the Greek plaintiff was able to methodically construct his case and the Greek mathematician his theorem. Nowhere else on Earth had men thought to agree on what constituted the essence of such and such a thing. Uniquely, the Greeks tore themselves away from the word of rulers, gods, or tradition, in favor of logical reasoning. What is wrongdoing? What is murder? What is theft? The ancient Greeks were the first to ask themselves these kinds of questions. They were the first to distinguish a "criminal act" from "misfortune" or "error of judgment." Definitions—concise, basic, and unambiguous—entered the Athenian imagination, as Aristotle tells us in *Rhetoric:*

> It often happens that a man will admit an act, but will not admit the accuser's label for the act... He will admit that he took a thing but not that he "stole" it; that he struck someone first, but not that he committed "outrage"; that he had intercourse with a woman, but not that he committed "adultery"; that he is guilty of "theft" but not of "sacrilege," the object stolen not being

> consecrated; that he has encroached, but not that he has "encroached on State lands"; that he has been in communication with the enemy, but not that he has been guilty of "treason". . . . Therefore we must be able to distinguish what is theft, outrage, or adultery, from what is not, if we are to be able to make the justice of our case clear . . .

Similarly, Euclid defines a "point," a "line," a "square," a "unit," and a "number" (among other things), as if answering questions that no one had thought to ask before. What is a point? What is a line? To a scribe in Alexandria, or a logician in Xianyang, these sentences were only puzzles. They had no meaning. Or else for response they would simply make a point or draw a line with a drop of ink.

Euclid's books not only posed these questions, they laid down the law (so far as the answers were concerned) for future generations of mathematicians. What is a point? That which has no part. A line? A length without breadth. A square? A quadrilateral, which is both equilateral and right-angled. A unit? That of which each of the things that exist is called one. A number? A multitude composed of units. Et cetera. These foundations allowed mathematicians to "be able to make the justice of [their] case clear." Not to mention the case of others. In the mid-nineteenth century, more than two millennia after Euclid, a copy of his *Elements* traveled in the carpetbag of a circuit lawyer from Illinois. His name was Abraham Lincoln.

The pages and their propositions made a deep impression

on Lincoln's mind, following him into his subsequent career in politics. In a speech given to an Ohio crowd in 1859 in opposition to a pro-slavery rival, one Judge Douglas, Lincoln declared, "There are two ways of establishing a proposition. One is by trying to demonstrate it upon reason, and the other is, to show that great men in former times have thought so and so, and thus to pass it by the weight of pure authority. Now, if Judge Douglas will demonstrate somehow that this is popular sovereignty,—the right of one man to make a slave of another, without any right in that other, or anyone else to object; demonstrate it as Euclid demonstrated propositions,—there is no objection. But when he comes forward, seeking to carry a principle by bringing it to the authority of men who themselves utterly repudiate that principle, I ask that he shall not be permitted to do it."

Definitions and axioms would shape President Lincoln's most famous addresses. His powers of rhetoric, persuasion, deduction, and logic were all subjected to the severest tests. The nation was in crisis. Civil war would shortly come. Still, the president spoke to the entire country in defense of the Union.

> I hold, that in contemplation of universal law, and of the Constitution, the Union of these States is perpetual. Perpetuity is implied, if not expressed, in the fundamental law of all national governments. It is safe to assert that no government proper ever had a provision in its organic law for its own termination. Continue to execute all the express provisions of our national

> Constitution and the Union will endure forever, it being impossible to destroy it except by some action not provided for in the instrument itself.

> Proposition: The Union of these States is perpetual.
>
> Clarification: Perpetuity is implied in the fundamental law of all national governments.
>
> Axiom: No government ever had a legal provision for its own termination.
>
> Conclusion: Therefore continue to execute the Constitution and the Union will endure forever.

Throughout Lincoln's four years in office, intense fighting saw approximately 750,000 men killed and the nation all but tear itself apart, but the president's proof would ultimately be vindicated.

"We are not enemies," the president had said in the same national address, "but friends." Perhaps he was thinking of a proverb attributed to Pythagoras, one that he took as an axiom: "Friendship is equality."

Eight

ON BIG NUMBERS

In the second of his *Olympian Odes,* the ancient lyric poet Pindar wrote, "the sand escapes numbering." He was expressing the same idea that would lead his fellow Greeks to coin the term "sand hundred" for an inconceivably great quantity.

Pindar's claim remained unassailable for some two centuries, which of course is not bad at all as far as a line of poetry goes. The eventual refutation, composed in the middle of the third century BCE, can be fairly listed among the finest achievements of the mathematician Archimedes.

Introducing his academic paper—the first in recorded history—to the king of his day, Archimedes made a spectacularly audacious argument.

> Some people believe, King Gelon, that the grains of sand are infinite in number. I mean not only the sand in Syracuse and the rest of Sicily, but also the sand in the whole inhabited land as well as the uninhabited. There are some who do not suppose that they are infinite, but that there is no number that has been named which is

> so large as to exceed its multitude . . . I will attempt to prove to you through geometrical demonstrations, which you will follow, that some of the numbers named by us . . . exceed . . . the number of grains of sand having a magnitude equal to the earth filled up.

Archimedes' estimations for the measurements of the Earth, moon, sun, and the other stars were generous: for example, making Earth's perimeter ten times larger than the calculations of earlier astronomers. Similarly, Archimedes went to great lengths to provide a capacious margin for error concerning the estimated size of a grain of sand. He compared ten thousand grains to the scale of a poppy seed, and then patiently lined the seeds end to end on a smooth ruler. In this way he measured the number of poppy seeds required to reach an inch as being twenty-five. This figure he adjusted still further, changing it to forty-seeds-per-inch length, so as to "prove indisputably what is proposed." Thus he calculated as sixteen million (10,000 × 40 × 40) the maximum number of grains of sand that could fill one square inch.

Archimedes assumed that the universe was spherical. He estimated a value for the diameter of the universe using calculations for the diameter of the Earth's orbit around the sun. The universe, according to his reckoning, had a diameter no greater than 100,000,000,000,000 stadia (about two light-years). 1,000,000,000,000,000,000,000,000,000,000,000,000,000,000,000,000,000,000,000,000,000,000,000,000,000 grains of sand would more than saturate the whole of space.

Next, Archimedes showed that the Greek term "myriad"

(ten thousand or a hundred hundreds) more than sufficed for the purpose of counting even the largest worldly quantities. The phrase "myriad myriads," he pointed out, allowed the counter to reach the equivalent of one hundred million—the largest named number in his time. But, he continued, if it is possible to count in myriads, it should be equally possible to count in "myriad myriads" so that multiplying the latter by itself the counter could attain "myriad myriad myriad myriads" or 10,000,000,000,000,000. And considering this new figure likewise as a unit, as respectable as a "myriad" or "myriad myriads," the counter could multiply "myriad myriad myriad myriads" by itself and proceed to: "myriad myriad myriad myriad myriad myriad myriad myriads" or 100,000,000,000,000,000,000,000,000,000,000.

Up till now we have multiplied a myriad by itself a total of eight times. Archimedes' next step possessed all the elegance of simple logic: multiply a myriad myriads by itself myriad myriads times over. The "1" that starts the resulting number is tailed by eight hundred million zeroes.

Doggedly pursuing his logic, Archimedes proposed multiplying this new number by itself up to as many as a myriad myriads times over: a number requiring the insertion of eighty quadrillion (80,000,000,000,000,000) zeroes after the one.

Archimedes concluded his paper in confident, if understated tones.

> King Gelon, to the many who have not also had a share of mathematics I suppose that these will not appear

> readily believable, but to those who have partaken of them and have thought deeply about the distances and sizes of the Earth and sun and moon and the universe this will be believable on the basis of demonstration. Hence, I thought that it is not inappropriate for you too to contemplate these things.

We find the same comparison between immensity and grains of sand in the sutras of India, many of which were set down on paper in Archimedes' time. In the *Lalitavistara* sutra, a hagiographical account of the Buddha's life, we read of a meeting between the young Siddhartha and the "great mathematician Arjuna." Arjuna asks the boy to multiply numbers a hundredfold beginning with one *koti* (generally considered the equivalent of ten million). Without the slightest hesitation, Siddhartha correctly replies that one hundred *koti*s equals an *ayuta* (which would equate to one billion), and then proceeds to multiply this number by one hundred, and the new number by one hundred, and so on, until—after twenty-three successive multiplications—he reaches the number called *tallaksana* (the equivalent of 1 followed by 53 zeroes).

Siddhartha proceeds to multiply this number in turn, though it is unclear whether he does so by one hundred or some other amount. In a phrase reminiscent of Archimedes, he claims that with this new number the mathematician could take every grain of sand in the river Ganges "as a subject of calculation and measure them." Again and again, the *bodhisattva* multiplies this number, until at last he reaches *sarvaniksepa,* with which, he tells the mathematician, it would be possible to count every grain of sand in ten rivers the size of the Ganges. And if this were not enough, he continues, we can multiply this number to reach *agrasara*—a number greater than the grains of sand in one billion Ganges.

Such extreme numerical altitudes, we are told, are the preserve of the pure and enlightened mind. According to the sutra, only the *bodhisattvas*, beings who have arrived at their ultimate incarnation, are capable of counting so high. In the closing verses, the mathematician Arjuna concedes this point.

> This supreme knowledge I do not have—he is above me.
> One with such knowledge of numbers is incomparable!

The story of the enlightenment of Siddhartha Gautama, to give him his full name, begins in his father's palace. It is said that the Nepalese king resolved to seclude his son at birth from the heartbreaking nature of the world. Shut up behind gilded doors, the boy would remain forever innocent of suffering, poverty, and death. We can imagine his constricted royal life: the fine meals of rich food, lessons in literature and military arts, ritual music and dance. In his ears he wore precious stones heavy enough to make his earlobes droop. But of course he was not free: he had only walls for a horizon, only ceilings for a sky. Bangle strings and brass flutes displaced all birdsong. Cloying aromas of cooked food overlay the smell of rain.

Nearly thirty years, a marriage, and even the birth of a son all passed before Siddhartha learned of a world beyond the palace walls. Having resolved to go forth and see it, he made a trip through the countryside, accompanied only by the charioteer who drove him. The prince saw for the first time men enfeebled by ill health, old age, and want of money. He was not even spared the sight of a corpse. Deeply shocked by all that he had seen, he fled his old life for the ascetic's road.

The story of the prince's seclusion in a palace reads like a fairy tale—it may very well be such a tale—with all its peculiar and thought-provoking charm. One particular aspect of Siddhartha's revelation of the outside world has always struck me. Quite possibly he lived his first thirty years without any knowledge of numbers.

How must he have felt, then, to see crowds of people mingling in the streets? Before that day he would not have believed that so many people existed in all the world. And what wonder it must have been to discover flocks of birds, and piles of stones, leaves on trees, and blades of grass! To suddenly realize that, his whole life long, he had been kept at arm's length from multiplicity.

Later, his followers would associate Siddhartha's enlightened mind with a profound knowledge of numbers. Perhaps as much as all the other surprises he witnessed from his chariot, it was this first sighting of multiplicity that set him on the path to Nirvana.

I am reminded of another story. This time the man was not a prince but a mathematician in mid-twentieth-century America. Big numbers fascinated him; he enjoyed talking about them with his nine-year-old nephew. One day, the mathematician, Edward Kasner, invited the boy to name a number that contains a hundred zeroes after the 1. "Googol," the boy replied, after a little thought.

No explanation for the origin of this word is given in Kasner's published account *Mathematics and the Imagination.* Probably, it came intuitively to the boy. According to linguists, English speakers tend to associate an initial G sound with the idea of bigness, since the language employs many G- words to describe things which are "great" or "grand," "gross" or "gargantuan," and which "grow" or "gain." I could point out another feature: both the elongated "oo" vowel and the concluding L suggest indefinite duration. We hear this difference in verbs like "put" and "pull," where "put"—with its final *T*—implies a completed action, whereas an individual might "pull" at something for any conceivable amount of time.

In a universe teeming with numbers, no physical quantity exists that coincides with a googol. A googol dwarfs the number of grains of sand in all the world. Collecting every letter of every word of every book ever published gets us nowhere near. The total number of elementary particles in all of known space falls some twenty zeroes short.

The boy could never hope to count every grain of sand, or read every page of every published book, but, like Archimedes and the Siddhartha of the sutras, he understood that no cosmos would ever contain all the numbers. He understood that with numbers he might imagine all that existed, all that had once existed or might one day exist, and all that existed too in the realms of speculation, fantasy, and dreams.

The mathematician liked his nephew's word. He immediately encouraged the boy to count higher still and watched as his small brow furrowed. Now came a second word, a variation of the first: "googolplex." The suffix *-plex* (duplex) parallels the English *-fold,* as in "tenfold" or "hundredfold." This number the boy defined as containing all the zeroes that a hand could write down before tiring. His uncle demurred. Endurance, he remarked, varied a great deal from person to person. In the end they agreed on the following definition: a googolplex is a 1 followed by a googol number of zeroes.

Let us pause a brief moment to contemplate this number's size. It is not, for instance, a googol times a googol: such a number would "only" consist of a 1 with 200 zeroes. A googolplex, on the other hand, contains far more than a thousand zeroes, or a myriad zeroes, or a million or billion zeroes. It contains far more than the eighty quadrillion zeroes at which even the painstaking and persistent Archimedes ceased to count. There are so many zeroes in this number that we could never finish writing them all down, even if every human lifetime devoted itself exclusively to the task.

The googolplex is so vast a number that it encompasses virtually every conceivable probability. Physicist Richard Crandall gives the example of a can of beer that spontaneously tips over, "an event made possible by fundamental quantum fluctuations," as having vastly greater than 1-in-a-googolplex odds. A further illustration, by the English mathematician John Littlewood, asks us to imagine the plight of a mouse in outer space. Littlewood calculated as being well within a googolplex the likelihood that this mouse—helped by sufficient random fluctuations in its environment—might survive a whole week on the surface of the sun.

But of course a googolplex is not infinite. We can, as perhaps the boy did, continue to count by simply adding one. Modern computers, impervious to zero vertigo, have calculated that this number, googolplex + 1, is not prime. Its smallest known factor is: 316,912,650,057,057,350,374,175,801,344,000,001.

What did the mathematician make of his nephew's "googolplex" as the biggest number he could conceive of? His reply is not recorded, but he might have told him about some of the infinitely many numbers that exceed the googolplex's scope. He might, for example, have mentioned "googol factorial," being the product of multiplying every whole number between 1 and a googol (1 × 2 × 3 × . . . 950,345 × . . . 1,000,000,000,000,008,761 × . . . googol). This number, which computers tell us begins 16294 . . . easily surpasses every other number that we have encountered in these pages.

For a universe of such limited dimensions, these monstrous numbers seem quite useless. Worse, they can appear to us excessive, disproportionate. Every number, after a certain point, feels gratuitous. Who knows? It is possible they are not intended for our attention. The Flower Adornment sutra speaks of immense aeons, *kalpas*, in which the universe is continuously destroyed and reborn. At the kalpa's peak, men live for an average of eighty-four thousand years. In other realms, so the *Heart* sutra reports, a single life spans eighty-four thousand *kalpas*—that is to say, eighty-four thousand epochs, each one many zeroes long. For such beings, a googol or its factorial would belong merely to the tangible and the convenient.

Mathematicians aspire to these heavenly realms. Vast numbers that split our senses enrich their work. But they also produce paradoxes. For instance, which is greater: 10 or 27, when each is multiplied by itself exactly a googolplex number of times? The latter, of course, although even the most powerful calculators—plunging one hundred digits deep—struggle to tell the two apart. This difficulty confounds our expectations: intuitively, we feel that the ordering of a number should remain straightforward, even when the number's precise value cannot be known. And yet, there exist numbers so large that we cannot easily distinguish them from their double, or triple, or quadruple or any other amount. There exist magnitudes so immense that they escape all our words, and all our numbers.

The most famous paradox concerning big numbers takes us back once more to the ancient Greeks. Tradition attributes it to the philosopher Eubulides. It has been suggested that Eubulides' inspiration owes something to his fellow skeptic Zeno, who argued that every falling grain of wheat makes a noise proportionate to the noise made by a falling bushel. Eubulides' formulation does not feature wheat, however. As would Archimedes a century later, Eubulides built his argument on sand.

It goes as follows: first, we agree that one grain of sand does not make a heap. Adding a second grain does not make a heap either. Nor do we produce a heap with the third grain. From this it follows that adding one to any small number generates another number we call "small." But if this is true, a billion is a small number. So, too, is a googolplex.

Understandably wary of this conclusion, the reader might propose that a heap of sand, like a big number, begins at a certain point: say, ten thousand. As a resolution of the paradox, this answer is unsatisfactory. It is unclear why nine thousand nine hundred and ninety-nine should be considered small, but not nine thousand nine hundred and ninety-nine plus one.

Of course, in at least one sense all numbers are small. Given any number "n," there will be only n-1 numbers less than n, but an infinite group of numbers greater than it.

Addressing the unjustly forgotten mathematician Archytas, the Roman poet Horace, who was regarded as the finest lyric poet of the era of Emperor Augustus, expressed in his verse perhaps the greatest paradox of them all: that of finite men who spend their lives attempting to scale the infinite.

> You who measured the sea and the earth and the
>   numberless sands,
> you, Archytas, are now confined in a small mound
>   of dirt
> near the Matine shore, and what good does it do you
>   that you
> attempted the mansions of the skies and that you
>   traversed
> the round celestial vault—you with a soul born to die?

Nine

SNOWMAN

Outside it is cold, cold—ten degrees below, give or take. I step out with my coat zipped up to my chin and my feet encased in heavy rubber boots. The glittering street is empty; the wool-gray sky is low. Under my scarf and gloves and thermals I can feel my pulse begin to make a racket. I do not care. I observe my breath. I wait.

A week before, not even a whole week, the roads showed black tire tracks and the trees' bare branches stood clean against blue sky. Now Ottawa is buried in snow. My friends' house is buried in snow. Chilling winds strafe the town.

The sight of falling flakes makes me shiver; it fills the space in my head that is devoted to wonder. How beautiful they are, I think. How beautiful are all these sticky and shiny fragments. When will they stop? In an hour? A day? A week? A month? There is no telling. Nobody can second-guess the snow.

The neighbors have not seen its like in a generation, they tell me. Shovels in hand, they dig paths from their garage doors out to the road. The older men affect expressions both of nonchalance and annoyance, but their expressions soon come undone. Faint smiles form at the corners of their wind-chapped mouths.

Granted, it is exhausting to trudge the snowy streets to the shops. Every leg muscle slips and tightens; every step forward seems to take an age. When I return, my friends ask me to help them clear the roof. I wobble up a leaning ladder and lend a hand. A strangely cheerful sense of futility lightens our labor: in the morning, we know, the roof will shine bright white again.

Hot under my onion layers of clothing, I carry a shirtful of perspiration back into the house. Wet socks unpeel like Band-Aids from my feet; the warm air smarts my skin. I wash and change my clothes.

Later, around a table, in the dusk of a candlelit supper, my friends and I exchange favorite recollections of winters past. We talk sleds, and toboggans, and fierce snowball fights. I recall a childhood memory, a memory from London: the first time I heard the sound of falling snow.

"What did it sound like?" the evening's host asks me.

"It sounded like someone slowly rubbing his hands together."

Frowns outline my friends' concentration. Yes, they say, laughing. Yes, we can hear what you mean.

A man sporting a gray mustache laughs louder than the others. I do not catch his name; he is not a regular guest. I gather he is some kind of scientist, of indeterminate discipline.

"Do you know why we see snow as white?" the scientist asks. We shake our heads.

"It is all to do with how the sides of the snowflakes reflect light." All the colors in the spectrum, he explains to us, scatter out from the snow in roughly equal proportions. This equal distribution of colors, we perceive as whiteness.

Now our host's wife has a question. The ladle with which she has been serving bowls of hot soup idles in the pot. "Could the colors never come out in a different proportion?" she asks.

"Sometimes, if the snow is very deep," he answers. In which case, the light that comes back to us can appear tinged with blue. "And sometimes a snowflake's structure will resemble that of a diamond," he continues. Light entering these flakes becomes so mangled as to dispense a rainbow of multicolored sparkles.

"Is it true that no two snowflakes are alike?" This question comes from the host's teenage daughter.

It is true. Imagine, he says, the complexity of a snowflake (and enthusiasm italicizes his word "complexity"). Every snowflake has a basic six-sided structure, but its spiraling descent through the air sculpts each hexagon in a unique way: the minutest variations in air temperature or moisture can—and do—make all the difference.

Like mathematicians who categorize every whole number into prime numbers or Fibonacci numbers or triangle numbers or square numbers (and so on) according to its properties, so researchers subdivide snowflakes into various groupings according to type. They classify the snowflakes by size and shape and symmetry. The main ways in which each vaporous hexagon forms and changes, it turns out, amount to several dozen or several score (the precise total depending on the classification scheme).

For example, some snowflakes are flat and have broad arms, resembling stars, so that meteorologists speak of "stellar plates," while those with deep ridges are called "sectored plates." Branchy flakes, like the ones seen in Christmas decorations, go by the term "stellar dendrites" (dendrite coming from the Greek word for a tree). When these tree-like flakes grow so many side branches that they finish by resembling ferns, they fall under the classification of "fern-like stellar dendrites."

Sometimes, snowflakes grow not thin but long, not flat but slender. They fall as columns of ice, the kinds that look like individual strands of a grandmother's white hair (these flakes are called "needles"). Some, like conjoined twins, show twelve sides instead of the usual six, while others—viewed up close—resemble bullets (the precise terms for them are "isolated bullet," "capped bullet," and "bullet rosette"). Other possible shapes include the "cup," the "sheath," and those resembling arrowheads (technically, "arrowhead twins").

We listen wordlessly to the scientist's explanations. Our rapt attention flatters him. As he speaks, his white hands draw the shape of every snowflake in the air.

Complexity. But from it, patterns, forms, identities, that every culture can perceive and understand. I have read, for instance, that the ancient Chinese called snowflakes blossoms and that the Scythians compared them to feathers. In the Psalms (147:16), snow is a "white fleece," while in parts of Africa it is likened to cotton. The Romans called snow *nix*, a homonym — the seventeenth-century mathematician and astronomer Johannes Kepler would later point out — of his Low German word for "nothing."

Kepler was the first scientist to describe snow. Not as flowers or fleece or feathers, snowflakes were at last perceived as being the product of complexity. The reason behind the snowflake's regular hexagonal shape was "not to be looked for in the material, for vapor is formless." Instead, Kepler suggested some dynamic organizing process, by which frozen water "globules" packed themselves together methodically in the most efficient way. "Here he was indebted to the English mathematician Thomas Harriot," reports the science writer Philip Ball, "who acted as a navigator for Walter Raleigh's voyages to the New World in 1584–5." Harriot had advised Raleigh concerning the "most efficient way to stack cannonballs on the ship's deck," prompting the mathematician "to theorize about the close packing of spheres." Kepler's conjecture that hexagonal packing "will be the tightest possible, so that in no other arrangement could more pellets be stuffed into the same container" would only be proven in 1998.

That night, the snow reached even into my dreams. My warm bed offered no protection from my childhood memories of the cold. I dreamed of a distant winter in my parents' garden: the powdery snow, freshly fallen, was like sugar to my younger brothers and sisters, who hastened out to it with shrieks of delight. I hesitated to join them, preferred to watch them playing from the safety of my bedroom window.

But later, after they had all wound up their games and headed back in, I ventured out alone and started to pack the snow together. Like the Inuit (who call it *igluksaq,* "house-building material"), I wanted to surround myself with it, build myself a shelter. The crunching snow gradually encircled me, accumulating on all sides, the walls rising ever higher until at last they covered me completely. My boyish face and hands smeared with snow, I crouched deep inside feeling sad and feeling safe.

"On t'attend!" my friends call up in the morning to my room. "We are ready and waiting!" I am the English slowpoke, unaccustomed to this freezing climate, the lethargy it imposes on the body, and the dogged, unshakable feeling of being underwater. What little snow I have experienced all these years, I realize now, has been but a pale imitation of the snow of my childhood. London's wet slush, quick to blacken, has muddied the memory. Yet here the Canadian snow is an irresistible, incandescent white—its glinting surfaces give me back my young days, and alongside them, a melancholic reminder of age.

After my sweater, I pull on a kind of thermal vest, then a coat, knee-long. My neck is wrapped in a scarf; my ears vanish behind furry muffs. Mitten fingers tie bootlaces into knots.

Fortunately, the Canadians have no fear of winter. The snow is well superintended here. Panic, of the kind that grips London or Washington, DC, is unfamiliar to them; stockpiling milk, bread, and canned food is unheard of. Traffic jams, canceled meetings, energy blackouts, are rare.

The faces that greet me downstairs are all kempt and smiling. They know that the roads will have been salted, that their letters and parcels will arrive on time, that the shops and schools will be open as usual for business.

In the schools of Ottawa, children extract snowflakes from white sheets of paper. They fold the crisp sheet to an oblong, and the oblong to a square, and the square to a right-angled triangle. With scissors, they snip the triangle on all sides; the pupils all fold and snip the paper in their own way. When they unravel the paper, different snowflakes appear, as many as there are children in the class. But every one has something in common: they are all symmetrical.

The paper snowflakes in the classroom resemble only partly those that fall outside the window. Shorn of nature's imperfections, the children's unfolded flakes represent an ideal. They are the pictures that we see whenever we close our eyes and think of a snowflake: equidistant arms identically pliable on six sides. We think of them as we think of stars, honeycombs, and flowers. We imagine snowflakes with the purity of a mathematician's mind.

At the University of Wisconsin, the mathematician David Griffeath has improved on the children's game by modeling snowflakes not with paper, but with a computer. In 2008, Griffeath and his colleague Janko Gravner, both specialists in "complex interacting systems with random dynamics," produced an algorithm that mimics the many physical principles that underlie how snowflakes form. The project proved slow and painstaking. It can take up to a day for the algorithm to perform the hundreds of thousands of calculations necessary for a single flake. Parameters were set and reset to make the simulations as lifelike as possible. But the end results were extraordinary. On the mathematicians' computer screen shimmered a galaxy of three-dimensional snowflakes—elaborate, finely ridged stellar dendrites and twelve-branched stars, needles, prisms, every known configuration, and others, resembling butterfly wings, that no one had identified before.

My friends take me on a trek through the nearby forest. The flakes are falling intermittently now; above our heads, patches of the sky show blue. Sunlight glistens on the hillocks of snow. We tread slowly, rhythmically, across the deep and shifting surfaces, which squirm and squeak under our boots.

Whenever snow falls, people look at things and suddenly see them. Lampposts and doorsteps and tree stumps and telephone lines take on a whole new character. We notice what they are, and not simply what they represent. Their curves, angles, repetitions, command our attention. Visitors to the forest stop and stare at the geometry of branches, of fences, of trisecting paths. They shake their heads in silent admiration.

A voice somewhere says the Hull River has frozen over. I disguise my excitement as a question. "Shall we go?" I ask my friends. For where there is ice, there will inevitably dance ice skaters, and where there are ice skaters, there will be laughter and lightheartedness, and stalls selling hot pastries and spiced wine. We go.

The frozen river brims with action: parkas pirouette, wet

dogs give chase, and customers line up at the concessions. The air smells of cinnamon. Everywhere, the snow is on people's lips: it serves as the icebreaker for every conversation. Nobody stands still as they are talking: they shift their weight from leg to leg, and stamp their feet, wiggle their noses and exaggerate their blinks.

The flakes fall heavier now. They whirl and rustle in the wind. Everyone seems in thrall to the tumbling snowflakes. Human noises evaporate; nobody moves. Nothing is indifferent to its touch. New worlds appear and disappear, leaving their prints upon our imagination. Snow comes to earth and forms snow benches, snow trees, snow cars, snowmen.

What would it be like, a world without snow? I cannot imagine such a place. It would be like a world devoid of numbers. Every snowflake, unique as every number, tells us something about complexity. Perhaps that is why we will never tire of its wonder.

Ten

INVISIBLE CITIES

"We wish to see ourselves translated into stone and plants, we want to take walks in ourselves when we stroll around these buildings and gardens." This, says Nietzsche, is the purpose of the city: to create space and structure in which a person might think. Ostentatious church buildings, he complained, inhibited free thought. He argued for the ideal of a "wide" and "expansive" city, expansive in every sense of the word.

I remember these words every time I fly to New York City, a place where tall buildings aspire to the sky. Long shadows in the shape of skyscrapers alight on yellow taxis and hot dog carts. The city's buildings are home to eight million human beings. Among them number some of the most creative people in the world. People come here from every land, and from every language, for what reason? Quite possibly, they come here in order to think.

But New Yorkers, like the rest of us, do not pay much attention to their surroundings, how the city incites and informs their thoughts. There are exceptions, of course, and I am not only talking about those who are newly arrived. I am talking about mathematicians, who are tourists in every place. What with its towering buildings, and its grid's rectilinear streets, and the intersections named after numerals (Ninety-Third and Fifth Avenue), New York City was made for mathematicians.

Planning a city, or dreaming about one, invites us to think by numbers, to borrow some of the mathematicians' delight. Architects of cities and of individual buildings divide and categorize the air. In this portion: morning traffic; in that section: jogging in the park. Up on this level: office computers; beneath it: a parking lot. The designers translate numbers into symmetry, into shape and order and livable form. Cities are the embodiments of numerical patterns that contain and direct our lives. But all cities are invisible to start with.

Before New York the city, there was New York the idea: a mere glint in the European settlers' eye. They christened the woodlands and rivers and the trails of native clans that they discovered, *Nieuw Amsterdam*. Many years later, following the War of Independence, the nascent settlement would serve briefly as the fledgling Union's capital. Intangible visions could now be rendered concrete.

A commission, formed in 1811, found in favor of a plan for the mass building of "straight-sided and right-angled houses." Avenues were laid, precisely one hundred feet wide, and numbered (beginning with the easternmost) from one to twelve. Running at right angles to the avenues were passages turned into symmetrical streets (sixty feet in width), each allotted a consecutive number between one and one hundred and fifty-five. Street names acted like compass points, shepherding strangers and the easily lost to their destination. The rigid geometry imposed order, efficient commerce, and cleanliness, but it also obliterated many of Manhattan Island's natural spaces. In the words of one commissioner, the grid system became "the day-dawn of our empire."

New York City is an exception, though. Not all cities find their territory; many remain forever orphans, existing only in their inventor's dreams. What follows is my attempt to sketch a brief history of these invisible cities.

Plato, in his *Laws,* gives a recipe for the ideal city. Like any recipe maker, he puts great store by the precision with which he delineates its ingredients. At various points in the text, he insists rather heavily on a particular number. No margin exists for approximation in the Platonic design. Neither is there any room for discussion, since for Plato, the self-evident quality of his city is "as plain as the fact that Crete is an island."

Judging by his laws, Plato really intended limits. Without a city, he argues, man would dwell in a "fearful, illimitable desert." Such a man would know nothing of art or science; worse, he would hardly know himself. A city's limits should be carefully demarcated, set neither too big nor too small, so that its citizens might, with time and sufficient effort, be capable of putting a name to every face. This, in Plato's judgment, would prevent the blight of war, which had struck down so many great cities of the past. He quotes with approval the poet Hesiod's praise of moderation: "the half is often greater than the whole."

Starting from the principle that "numbers in their divisions and complexities are useful for everything," Plato proposes limiting his ideal city to exactly 5,040 landholding families. Why 5,040? It is what mathematicians call a "highly composite number," meaning that it can be divided in multiple ways. In fact, no fewer than sixty numbers divide into it, including every number from one to ten.

Twelve can also divide evenly into 5,040. Plato divides the total number of households into twelve tribes, each therefore consisting of 420 families. While interdependent, each tribe is conceived as being fixed and self-sufficient, like the months in a solar year.

Using highly composite numbers facilitates the subdivision of land and property among the citizens. Each family, in each tribe, receives an equal lot of land, beginning from the city center and radiating out to the countryside. In this way the city fairly distributes the fertility of its soil: half of each lot shall contain the city's richest ground, while the other half shall consist of the rockiest.

Plato's ideal number of 5,040 families intrigues modern statisticians. They calculate that such a population would require 164 (or 165) births per year to sustain itself. Following the ancient Greek logic that treated men as the head of every household, they also calculate the city's annual number of potential fathers to be 1,193. Plato believed that one marriage in seven would be fruitful per year, suggesting an anticipated annual birthrate of 170—which corresponds almost exactly to our statisticians' calculations.

How did Plato count on keeping his ideal city's number of households in check? He proposed that each inheritance should pass into the hands of a single "best-loved" male heir. Any remaining sons would be distributed among the childless citizens, while the daughters should be married off.

Large families had no place in Plato's city. Fecundity would be illegal: any couple producing "too many" children was to be rebuked. The city's precise limit of 5,040 households was inviolable: all surplus members were to be sent packing.

Plato imagined that his limits would ensure equality and security for every citizen. In his bucolic vision, men and women would "feed on barley and wheat, baking the wheat and kneading the flour, making noble puddings and loaves; these they will serve up on a mat of reeds or clean leaves; themselves reclining the while upon beds of yew and myrtle boughs. And they and their children will feast, drinking of the wine which they have made, wearing garlands on their heads, and having the praises of the gods on their lips, living in sweet society, and having a care that their families do not exceed their means; for they will have an eye to poverty or war."

Perhaps. But it is also conceivable that Plato's city would have encouraged just the kind of petty-mindedness among its citizens that exact calculation often fosters.

During the Renaissance, when Plato and his ideas were rediscovered by humanist scholars, we find an Italian architect similarly moved to envision his own perfect city. His name was Antonio di Pietro Averlino, though he is better known today by the Greek name Filarete (meaning "lover of virtue"). Unlike Plato, Filarete was an architect, albeit one with a rich and complicated past: he had once been arrested and barred from working in Rome for allegedly stealing the head of Saint John the Baptist.

Filarete described his city *Sforzinda* (the name a flattery aimed at his patron Francesco Sforza of Milan) at some length in his *Trattato di architettura*. Its thick symmetrical outer walls formed an eight-pointed star. Though attractive, the unusual choice of shape was also intended to be defensive: invaders mounting its angles would find themselves exposed on multiple sides.

Like spokes in a vast wheel, eight straight roads led from the walls to the city center. The roads were studded with small piazzas, surrounded by shops and markets. A visitor treading his path downtown from the city gates would pass pyramids of apples, stacks of loaves, and multicolored garments spilling over tables. Merchants, their eyes enlarged by expectation, shouting: *"Signore, signore!"* At last the city center appears. Three vast interconnecting piazzas greet him. Here the market noises fade before the imposing ducal palace to his left, and a massive cathedral on his right. In between the palace and the cathedral, on the main piazza, stands another lofty building, ten stories high.

What is this strange building to which every street in the city addresses itself? Filarete called it the "House of Vice and Virtue." Every floor housed a different class of activity.

A brothel, on the ground floor, would entertain the majority of the building's callers. Alcoholic drinks and games could be had on the floors immediately above. Ascending further, a university and lecture halls offered its few visitors instruction. An observatory topped them all.

The homes to which the citizens would return after their day's work or play had been just as intricately imagined. Filarete planned his houses according to the resident's social rank: the artisans' quarters taking up far less space than the houses of the city's merchants or gentlemen. In comparison with his artist and painter neighbors, the architect's own house would be twice as spacious.

Filarete's plans are long, his writing spidery. Spread across its twenty-five volumes his treatise contains an entire city in waiting. But not long after its completion, in 1466, the Duke Sforza died and Filarete's vision survived only on paper.

Plato's ideal limits and Filarete's volumes appealed to new dreamers. They passed from mind to mind, and from century to century. It should perhaps come as no surprise that they would go on to inspire the grandest and most ambitious city plans, those drawn up in the United States.

King Camp Gillette, who would later invent the safety razor, once dreamed of an immense city he called Metropolis. In 1894, he published a short illustrated book aimed at its promotion. The city, Gillette wrote, would be "situated in the vicinity of Niagara Falls, extending east into New York State and west into Ontario." It would take the shape of a rectangle, sixty miles long and thirty miles wide. Gillette considered its construction, "In the light of a machine, or rather a part of the machine of production and distribution; and, as such, the objects to be attained must be known and understood. It must have no unnecessary parts to cause friction or demand unnecessary labor, and yet it must combine within itself all the necessary parts which will contribute to the happiness and comfort of all."

Sixty million inhabitants would live in circular skyscrapers, six hundred feet in diameter, "upon a scale of magnificence such as no civilization has ever known." Elevators, a still recent invention in Gillette's day, had gradually shifted city design from the horizontal toward the vertical. But Metropolis took the idea of a vertical city to a whole other level. The skyscrapers would be truly colossal, attaining a height of twenty-five stories. Habitable monoliths, lots and lots of them, shining with glass and progress, smooth and steel-colored, a monotony of monuments. A beehive-like distribution of apartments throughout the city would lend space enough between the buildings for wide avenues and parks. No citizen, it also ensured, would reside any nearer or farther away from a school, shop, or theater than any other.

The city's highly regular order appears again in miniature in Gillette's plan for every home. For Gillette, as for Filarete, the home was a tiny city. The interior would be completely symmetrical, with parallel sitting rooms, bedrooms, and bathrooms on either side. And in each room, the windows would be so arranged as to make it impossible to look out from one apartment into a neighbor's.

Conscious of the artificiality of his vision (even the

hexagonal lawns that surrounded each building would be composed of artificial grass), Gillette proposed distributing thousands of public gardens filled with trees and "urns of flowers" at regular intervals throughout the city. Being constructed with complete regularity did not mean that it would suffer from sameness, he insisted. Looking out from his window, the citizen's roving eye would encounter "a continuous and perfectly finished facade from every point of view, each building and avenue surrounded and bordered by an ever changing beauty in flowers and foliage."

Gillette summed up his utopian vision:

> Imagine for a moment these thirty-odd thousand buildings of *Metropolis,* each standing alone, a majestic world of art... a never-ending city of beauty and cleanliness, and then compare it with our cities of filth, crime, and misery, with their ill-paved and dirty thoroughfares, crowded with the struggling masses of humanity and the system of necessary traffic. And then compare the machinery of both systems, and take your choice; for I believe the only obstacle that lies in the way of the building of this great city is man.

Fifty years after Gillette's book was published, the 1939 World's Fair in New York exhibited its own "City of the Future."

Millions stood in line for hours to see the model Democracity. (This fact alone is truly remarkable: New Yorkers hate to stand in line. They hate the involuntary intimacy, and the excruciating foot shuffles, and the boredom of being so long in their own company. And yet wait in line they did.)

How long must it have taken the model makers to build? It was housed inside a sphere eighteen stories tall. An escalator, the world's longest, ascended visitors fifty feet above the ground to the exhibition. Music blared triumphant over loudspeakers as they entered, and then a resounding voice spoke. "The city of man in the world of tomorrow. Here are grass and tree as well as stone and steel. Not a dream city but a symbol of life as lived by the man of tomorrow. As man helps man so nation leans on nation, united by a thousand roads of commerce... Here brain and brawn, faith and courage, are interlinked in high endeavor."

Standing on slowly rotating balconies, visitors peered down at the city as though at an altitude of seven thousand feet. They saw a vast ring, brightly lit and painted, representing eleven thousand square miles of terrain. In its center loomed a single block: an awesome office complex to which a quarter of a million inhabitants (one citizen in six) would commute every day.

Five satellite suburbs surrounded this central district in concentric circles. Even the most distant Pleasantville (as the residential areas were called) lay within sixty miles of the hub. Larger Millvilles, home to the city's factories, exiled their noise and pollution to the outskirts. Greenbelts were interspersed with suburbs; wide highways communicated between the center and the various sectors.

Mobility, an American obsession, had never been so well catered to. Traffic lights would be a thing of the past. The city's highways would always flow freely, in a straight line, planned in such a way as to avoid all jams and pedestrian crossings. All other roads would be built a safe distance from any school.

Two minutes into the performance, all of a sudden the lights would dim; the concave ceiling sparkled as if with stars. A chorus started up, and a film showed marching men: artisans, farmers, mechanics, everyone who would work together to help build tomorrow. The voices rose; the marching figures grew; the fairgoers held their breath.

And then, just as quickly, the music died away and the men in the film disappeared behind thick plumes of smoke. The show was over.

Eleven

ARE WE ALONE?

Democritus, a contemporary of Plato and Aristotle, imagined all matter as composed of indivisible elements he called "atomos," and was also the first thinker to propose a cosmos of many worlds. Every world was different. Some had neither a sun nor a moon, while others had larger moons or smaller moons or moons more numerous than our own.

The Pythagoreans also believed that our world was simply one of many. For them the moon was Earth-like and inhabited, containing larger beings and more beautiful plants. The lunar residents, they believed, stood fifty times our height, lived on air and for this reason produced no excrement.

Refutations of these ideas by both Plato (to call the number of worlds indefinite requires definite knowledge) and Aristotle did not prevent later thinkers from endorsing them.

> Since space lies empty and infinite in all directions and since atoms in countless numbers fly every which way through its furthest reaches . . . it is utterly unrealistic to

> think that ours are the only world and sky to have been born and that so many atoms outside our world are doing nothing . . . there are other worlds in other parts of the universe, different races of humans and species of animals.

These lines come from the epic poem *On the Nature of Things,* written by the Roman Lucretius in the first century before the Christian era. The poet's ideas would later fill the fathers of the early Church with consternation. If other worlds existed, wrote St. Augustine, they would each require their own Savior, which would contradict the unique role of Christ.

But by the medieval period, not everyone in the Church shared Augustine's view. In 1277, the Bishop of Paris denounced the proposition that God could not create more than one world. Three centuries later, the friar Giordano Bruno put forth an elaborate argument in favor of an infinite number of worlds: if man can imagine so many worlds, then so can God, who creates what He thinks. The friar pictured Gardens of Eden, infinite in number: in half, Adam and Eve eat the fruit of Knowledge; in half, they do not. An infinite number of worlds will fall from grace and require an infinite number of Saviors to redeem them. Unlike Augustine, Bruno had no difficulty imagining an infinite number of Christs. For this and other "theological errors," the authorities denounced the heretic and burned him at the stake.

The Inquisitors' heavy footsteps dissuaded Bruno's contemporary, Galileo Galilei, from seeing any evidence for extraterrestrial life in the rugged lunar landscape revealed by his telescope. All the same, since the valleys and mountains that corrugated the moon's surface seemed at least comparable to those on Earth, might not the moon also have people to dwell among them? His friend Johannes Kepler, the seventeenth-century mathematician and astronomer, thought so. Jupiter, too, he deduced "with the highest degree of probability," though its inhabitants were undoubtedly inferior to humans.

Probability: this word became the cornerstone of the argument for life on other planets. "As for mind beyond the confines of our tiny globe," wrote the American astronomer Percival Lowell in 1895, "modesty, backed by a probability little short of demonstration, forbids the thought that we are the sole thinkers in this great universe."

His argument, two millennia old, tallied with the most modern scientific observations: the conditions on Mars suggested hospitality. The planet enjoyed an atmosphere and its weather appeared extremely clement (with an average climate, he reported, that was similar to that in southern England). Water, essential for life, was also present. "Anyone looking through a telescope at the planet, early last summer," noted Lowell, "would at once have been struck by the fact that its surface was diversified by markings in three colors, white, blue-green, and reddish-ocher; the white lying in a great oval at the top of the disc. The white oval was the south polar icecap." And the blue-green? The color of water. Or, of what still remained of Martian water, since "the signs [are] that its water supply is now exceedingly low." Its inhabitants, he reasoned, had poured all their energies into irrigation. When he scanned the planet's surface, he picked out a "network of fine, straight dark lines." Canals. "All this," the astronomer conceded, "of course, may . . . signify nothing; but the probability seems the other way . . . that Mars seems to be inhabited is not the last, but the first word on the subject."

Lowell's claims found many sympathetic ears. "Probability," he knew, was a sesame word: it opened ears and minds. But its magic did not work on everyone. The biologist Alfred Russel Wallace, who independently of Darwin discovered the principle of natural selection, was among his fiercest critics. Yes, Mars appeared to have polar icecaps, and a day only half an hour longer than our own, and lengthy seasons that vanished one into the next. But according to Wallace's calculations, the planet was in fact too cold to have rivers, seas, or canals. The features observed by Lowell were natural landforms, all products of normal geological processes. "Mars, therefore, is not only uninhabitable by intelligent beings . . . but is absolutely uninhabitable."

Not only was the planet Mars lifeless, but there were in all likelihood no habitations on any other planet either. This, at the turn of the twentieth century, was Wallace's stark assessment. The exceptional (and exceptionally complex) combination and sequence of events—physical, chemical, cosmological—permitting the origin of life on Earth made the prospect of finding other beings elsewhere in the universe immeasurably remote. The formation of intelligent life, quite simply, was a once-in-a-universe event.

What? *Alone?* Many people could not believe it. It was

one thing to be alone in a room, or a house. But to imagine being the only person in a town, a city, a country! The only people in a universe. They agreed with the ancient Greek Metrodorus, who had thought it absurd that in a vast field only one stalk should ever sprout. What is more, the sense of such all-encompassing solitude seemed to these people intolerably oppressive. They felt like foreigners in a lifeless void.

Decades passed, and the only extraterrestrials were to be found in pulp stories and motion pictures. When the American astronomer Frank Drake—who has spearheaded research into extraterrestrial communication—was a boy, he drew on this imagery as he listened to the words of his father, an engineer. Look at this star, and this one, and that star, and that one. All these stars, innumerable stars, shining in the night sky. Around some of those stars, somewhere in space, circled other worlds like our own. The boy listened to his father and believed him. He believed with all his heart.

Frank Drake was at home with big numbers. His formative years were spent in Chicago, a city so populous that it could sustain the livelihoods of more than a hundred full-time piano tuners. His thoughts ran often to the many, many worlds far above his head. He wondered about their cities, about their cars, about whether they knew war or cancer.

After graduating from Harvard with a doctorate in radio astronomy, Drake conducted the first ever search for interstellar communication. On April 8, 1960, he aimed the radio at two stars much like our sun, twelve light-years from Earth. Over the next two months he and his colleagues listened sedulously for a signal, but heard nothing. Not a peep.

But the numbers! Drake believed the numbers were on his side. The number of stars in our galaxy amounts to at least one hundred billion. One hundred billion! And how many of these stars were suns to planets? No hard facts could lend a hand. His imagination groped, sifted, hesitated. He closed his eyes and made a guess: one in two. Planets would orbit half of all the stars in the Milky Way, which equals fifty billion solar systems.

Not every solar system will produce life, however. A shot at life requires a certain kind of solar system (a sun that is neither too cool nor too dim, nor so massive as to burn out before life appears), which would host planets in some way comparable to Earth. Drake thought of the only solar system with which we are familiar, of its (then) nine planets, and the sole planet that had become a world. This figure—one—troubled him: it smacked of uniqueness. No, no, there had to be solar systems with multiple worlds. Some system will host a world, then it will host another, and perhaps another after that. Why not? Look at Mars; it had not been so very far off from being a second Earth. And so he made the guess of two (two being greater than one) for the number of possible Earths in each solar system.

So far, so good. But Drake's next estimations demanded all his powers of invention. First, he had to estimate the number of these planets on which life has emerged. This is how he reasoned about it. Four and a half billion years ago, not long after our Earth was formed, it struck out as a cold and barren rock. From this unpromising start, a few hundred million years later, the first living cells arose. What are a few hundred million years in a universe of ten or twenty billion years? It is as though life will get going just as soon as it is given the chance. Life had come quite easily to Earth, he concluded, therefore it should come easily for all the other possible worlds.

Second, Drake pondered the question of intelligence. On how many of the hundred billion Earth-like planets might the living cells contrive intelligent forms? Earth's diversity presented itself to his mind. Billions of animal species had wriggled, and buzzed, and hissed, and flitted, and swum through the ages. Yet only one had ever posed itself questions, and dreamed of life on other worlds. Furthermore, *Homo sapiens* was a latecomer to a planet that had done without its brains for billions of years. It suggested to Drake that questioning minds were anything but universal. He finally settled upon a figure of one in a hundred, leaving him a billion potential civilizations amid the stars.

How many of these civilizations had developed the technology (not to mention, had the desire) to communicate with others and in ways that we could understand? Drake was a radio astronomer. He knew that transmissions from Earth had already seeped far into space. Out there somewhere, twenty or thirty light-years away, hands and ears with the know-how were already perhaps tuning in to the same episodes of *Flash Gordon* and *The Lone Ranger* that he had listened to as a boy. And some of these planets would certainly be broadcasting signals of their own: say, one hundred thousand in all.

These planets would exude music, news broadcasts, coded

messages, provided of course that they were still around, that this same technology had not yet blown them to pieces. Civilization, after all, is a tricky business. Hazardous, too. The civilized human dates back only ten thousand years, and already Washington, DC, had nuclear arms aimed at Moscow (and vice versa). The threat of mutual destruction, Drake knew, was no delusion. But neither was such a trajectory inevitable. Mindful of his final proviso, he set his expectations low: of the potential hundred thousand communicating civilizations, the survivors in our galaxy would number maybe ten. Signals, centuries and millennia old, from thousands of their predecessors would likewise be filtering through space, ready to brush some awaiting antenna.

Drake wrote out his reasoning in the impressive-looking shorthand of an equation.

$$N = N^* \times f^p \times n^e \times f^l \times f^i \times f^c \times fL$$

Where N is the number of communicating civilizations in our galaxy.

N^* is the number of stars in the Milky Way.

F^p is the fraction of stars with planets orbiting them.

N^e is the number of planets per star ecologically capable of life.

F^l is the fraction of those planets where life evolves.

F^i is the fraction of these living planets that evolves intelligent life.

F^c is the fraction of these that communicate with others.

fL is the fraction of a planet's life during which the civilization survives.

Now to a journalist's question, "Do other intelligent civilizations exist?" came Drake's reply: "With almost absolute certainty." A certainty identical to and inherited from the one he had heard in his father's voice.

What would the astronomer give to actually find another world! (Russian colleagues reminded him that "world" in their language is a homonym for "peace.") Every day, in his observatory, he set to work. Through thick glasses he followed the chart recorder's progress, watched the needle fidget, saw the ink illustrate the contours of random noise. Occasionally, growing more and more impatient, he would snatch up headphones and listen in to the reception. He sat stock-still in the room, his heart in his mouth, and listened. Listened for what exactly? For a beep, a buzz, an electronic whisper. He watched and listened and waited and the hours turned. But the surprise did not come. Nothing came, except the months, the years, the hours.

With time, the technology became refined and upgraded. More and more assistants lent their ears and patience to the expanding project. "Probability" was on everyone's lips. The numbers are on our side, they told the press, their friends, themselves. It is only a matter of time.

Queasy silence was all that they heard.

The radio telescopes grew, and with them the doubts.

You would have to be superhuman not to doubt. Except perhaps for Drake, who had invested so heavily in his hope.

The problem posed by the silence was striking, though. For if thousands of communicating civilizations have evolved throughout our galaxy over millions (or even billions) of years, why has not a single one colonized its environs, including Earth? A civilization far older than our own, out of the thousands predicted by Drake's equation, would need only a few million years — a cosmic blink of an eye — to have colonized the Milky Way. Or, at the very least, to have flooded us with signs of its presence.

Look longer! Listen harder! Drake's response was emphatic. Perhaps the other civilizations are checking us out before they get in touch. Or perhaps they content themselves with colonizing only their solar systems. Or perhaps the cost of interstellar travel is too steep. Or perhaps, they never invented the radio. Perhaps, perhaps, perhaps.

Is Anyone Out There? To coincide, in 1992, with the five-hundredth anniversary of Columbus's discovery of America, Drake published a book posing this question. The answer, he felt, was closer than ever. He wanted to "prepare thinking adults" for the "imminent detection of signals from an extraterrestrial civilization." What kind of civilization? They would resemble humans, with heads on the top of their bodies and two legs to walk on. But instead of two arms, they would have four: "four make for a much better design." They would also be immortal, innocent of death. "This discovery, which I fully expect to witness before the year 2000, will profoundly change the world."

In the same year, NASA's computers performed fifty million tests per second on the data from the largest ever and most sophisticated radio scan of the heavens. They found nothing.

Biologists, meanwhile, were offering a reassessment of the equation's assumptions. Drake and his colleagues applied "strictly deterministic thinking," wrote the Harvard biologist Ernst Mayr. "Such thinking is often quite legitimate for physical phenomena, but is quite inappropriate for evolutionary events or social processes such as the origin of civilizations." Another biologist, Leonard Ornstein, pointed out that "even if we allow that the universe may be awash with planets with flourishing 'protometabolism,' and even 'protocells,' it does not necessarily follow that contingent events that contributed to the next step in the origin of life have been mimicked on even one of these hypothetical worlds."

Ornstein suggested an analogy: imagine that we dipped our hand only once into a bag of marbles and withdrew just one marble. This marble is blue-green. Conclusion? We could as equally suppose that the bag contained only one-in-a-million other blue-green marbles, or none at all, as suppose that all or many still in the bag would possess a similar color.

The only thing we can know for certain is that the probability of intelligent life in our universe is above zero (for were it zero, I would not be here to write this sentence and you would not be here to read it). The rest is speculation. Since the big bang, there have been billions of civilizations.

There have been millions of civilizations. There have been thousands of civilizations. There have been hundreds of civilizations, or tens. Or just one.

Why not? Probability is often expressed using large but finite numbers: "one in a thousand," "one in a million." But perhaps the probability of life, intelligent life, appearing somewhere in our universe is *infinitesimal*. If so, a universe would need infinitely many planets to produce even a finite number of civilizations (i.e., one).

Such a conclusion ought to be at least as motivating to us as Drake's, especially in an age of high-stakes international diplomacy, atomic bombs, and climate change. As the astronomer Michael Papagiannis concluded, "Knowing we are the only ones might make us realize that we are too valuable to destroy."

Twelve

THE CALENDAR OF OMAR KHAYYÁM

For the Bedouins living before the era of Muhammad, time did not exist. Or rather, they thought of it as an all-enveloping and all-enfeebling mist, without clear shape or pattern. Only the bright stars overhead pricked the pervasive gloom, helping the nomads to anticipate rain and decide when to take their animals to pasture. To make sense of their lives, the men sang songs, and the songs they sang told of distant earthquakes and battles. It was the sole history that they knew.

The Prophet's birth, to believe tradition, coincided with one of these battles: the so-called Event of the Elephant (occurring in the latter half of the sixth century of the Christian Era), when Mecca fell under the siege of a foreign king's army with a white elephant at its helm. According to a tale later told in the Koran, God sent a cloud of birds to pelt the attackers with stones until they fled.

Alongside Muhammad's revelation of a new religion came his revelation of time. Gone now was the idea of life as constituted by a flux of vague, discontinuous, and casual moments. Five compulsory prayers—*Fajr* (at dawn), *Dhuhr* (after the sun's zenith), *Asr* (during late afternoon), *Maghrib* (at sunset), and *Isha* (at twilight)—regulated each day. All our days, said the Prophet, are numbered. Each follows the other in meaningful succession.

"God wraps night around day, and He wraps day around night."

Seven of them together make a week (beginning on what we call Saturday), the span in which, it was said, God had progressively created the world: the earth on the first day, the hills on the second, the trees on the third, all unpleasant things on the fourth, the light on the fifth, the beasts on the sixth, and Adam, who was the last of creation, about the time of the *Asr* prayer on the seventh.

Look up at the heavens, Muhammad urged his followers. Each month, he declared, began when the moon appeared "like an old shriveled palm-leaf." Divinely ordained properties separated the months and made them distinct. During four of the months, it was forbidden to draw a sword. During certain others, believers could set out on pilgrimage. One month, called Ramadan, was set aside for fasting during the hours of daylight. Twelve lunar months composed one year.

Muhammad had been preaching for about a decade when, at the approximate age of fifty, he, along with his small band of followers, was ousted from the city by the rulers of Mecca. On camels, they traveled north to the oasis town of Yathrib, finding refuge there. The flight, known as *hegira,* became the founding date of the Islamic calendar; henceforth, every period of time would be precisely accounted.

Exquisitely contrived clocks sprang up across the Islamic world throughout the medieval period. The most impressive of these their makers filled not with sand, but water. A Chinese traveler's report from a visit to Antioch, three centuries after Muhammad, tells of a water clock within the royal residence that suspended twelve golden balls corresponding to the twelve hours of the day. At each hour one of the balls fell, splashing two o'clock, three o'clock, four o'clock, "the sound of which makes known the divisions of the day without the slightest error." Another water clock, described by Al-Jazari in his 1206 *Book of Knowledge of Ingenious Mechanical Devices,* stood as high as two men and featured robotic drummers, trumpeters, and cymbalists who played in turns according to the time of day.

By employing water—a precious substance in a peninsula largely composed of desert—the clockmakers demonstrated the reverence they accorded to time. Indeed, the Prophet had told his faithful to pay minute attention to its reckoning. Mosques employed *muwaqqits* (timekeepers) to calculate the official hours for each prayer. Generations of scholars scrupulously debated the world's age. The historian Al-Tabari, for example, figured that the world would endure in total seven thousand years, of which his generation had before them only two hundred. He based his calculation on a saying of the Prophet, in which Muhammad likens the time remaining until the Last Day to the fraction of time between the *Asr* (afternoon) prayer and sunset (about one-fourteenth of a day).

Writing in the fourth and fifth centuries of *hegira* (the eleventh century according to our way of reckoning), a near contemporary of Al-Tabari, Al-Biruni, compiled his *Chronology of Ancient Nations.* In it, he compares the calendars of the other great civilizations. The Greeks, the Syrians, and the Egyptians, he noted, all used a calendar of 365 and one one-quarter days, summing the quarters to make one complete extra day every four years.

"The ancient Egyptians followed the same practice, but with this difference, that they neglected the quarters of a day till they had summed up to the number of days in one complete year, which took place every 1,460 years; then they added one extra year."

The ancient Persians, he continued, also neglected these quarters of a day, though for a period of 120 years, after which they added one extra month.

As fate would have it (though Muslims believe no such thing exists, and that every moment is the conscious creation of God), Al-Biruni died in the same year that a boy was born to a Persian tentmaker. The Farsi word for tentmaker is *khayyám;* the tentmaker named his son Omar.

It is probable that as a child he studied the Koran. He would have learned to recite its verses aloud, for tradition holds that the scripture is akin to a chant, which is why the angel Gabriel chose to speak its words to the illiterate Muhammad. Perhaps the boy recited such a verse as, "Most surely in the creation of the heavens and the earth, and the alteration of the night and the day, there are signs for men who understand."

Many other books, on many subjects, must also have passed through his hands: books on geometry and the movement of the stars, books on arithmetic and music. He would have learned many of the pages by heart. It is likely that he also read or heard of Al-Biruni's compendium of calendars, and smiled at Al-Tabari's apocalyptic prediction. From long years of cloistered study, indifferent to the company of other people, he earned the bookish reputation as a "bad character."

One story relates how a student visited Khayyám daily before dawn to learn from him, before badmouthing him to the other townspeople. Hearing of this, Khayyám secretly invited the town's musicians to call on him at dawn the next day. When the unsuspecting student arrived as usual for his lesson, Khayyám ordered the musicians to beat their drums and blow their trumpets; the commotion awakened people from every quarter. "Men of Nishapur," Khayyám said, addressing the crowd, "he comes every day at this hour to my house, and studies with me, but to you he speaks of me in the manner you know. If I am really as he says, then why does he come and study with me? And if not, why does he abuse his teacher?"

When he was not reading books, he wrote them. A gifted poet, he was better known in his day as a talented mathematician. "The notion that one could use geometric constructions for certain types of algebraic problems was certainly recognized by Euclid and Archimedes," writes the mathematician

Ramesh Gangolli, "but before Omar Khayyám's construction, only simple types of equations... were thought to be amenable to the geometric method... Khayyám opened the door to the study of the more general question: What kind of algebraic problems can be represented and solved successfully in this manner?"

The young Persian's receptivity to inspiration must have been immense. When the sunlight shone through the latticework windows of his study, it danced upon the walls in geometrical shapes. Khayyám's pen traced *rubai* (poems) of four short rhyming lines, tight as theorems, writing the words from right to left. Some say he composed only sixty such poems; others, six hundred. He also wrote a commentary on Euclid's *Elements* that Gangolli tells us explained "in more detail many aspects that were left implicit and clarified many misconceptions about the structure of axiomatic systems."

Polyvalent talent like his is rare in any age. It likely led to jealousies, snide comments, upturned noses from certain quarters among his fellow countrymen. In one of several treatises on algebraic problems, Khayyám complained about the trials of the mathematician's life.

> I was unable to devote myself to the learning of this algebra and the continued concentration upon it, because of obstacles... which hindered me; for we have been deprived of all the people of knowledge save for a group, small in number, with many troubles, whose concern in life is to snatch the opportunity, when time

> is asleep, to devote themselves meanwhile to the investigation and perfection of a science; for the majority of people who imitate philosophers confuse the true with the false, and they do nothing but deceive and pretend knowledge, and they do not use what they know of the sciences except for base and material purposes.

In 452 (1074 CE by the Julian calendar), Sultan Jalal Al-Din invited Omar to the capital. His long Farsi texts, packed with numbers and equations, had preceded him. Even at the zenith of the golden age of Islamic mathematics, Khayyám's talent marked him out. Anxiously, expectantly, he must have followed his guide through the halls of the turquoise-studded palace. Tessellating tiles ran the length of the floor; mirrors hanging on the walls returned to him every one of his features, down to the wrinkles of laughter around his eyes.

The Sultan that Khayyám met hardly looked the part: he was very young, not yet twenty years of age, and keen to make his mark. His illustrious guest, the Sultan's Vizier reports, was promptly showered with the prince's praises, and offered an annual pension of 1,200 *mithkáls* of gold. For this, Khayyám agreed to accept an important commission: he was to create—provided that God were willing—a new civil calendar in the young Sultan's name.

Persia, geographically vast and intricately governed, had long lived on two times: while the imams practiced the faith using the *hegira* calendar, the bureaucrats counted their days by looking to the sun. The old civil calendar consisted of twelve months each of thirty days, excepting the eighth month, which contained thirty-five days. Its total of 365 days (eleven days more than in a lunar year) helped fix administrative dates far more closely to the seasons: an essential requirement when the nation's annual tax revenues relied heavily on the autumn harvests. But even these eleven extra days could not keep up precisely with the seasons: each year, as Al-Biruni had recorded, accumulated a lag of a quarter day. This was the problem that Khayyám set out to solve.

Day and night, Omar pondered how best to reform the old civil calendar. Astronomy had never before found itself so flattered and moneyed. A large observatory duly arose, from which Khayyám and his colleagues stalked the sky. He studied the sun's pathway through the twelve constellations of stars, and compiled detailed statistical tables. With this data he mapped a calendar based on the seasons, beginning the year (which the Persians called *Nowruz*) on the spring equinox (March 20 or 21), with the fourth month falling on the summer solstice (June 21), the seventh month on the autumn equinox (September 21), and the tenth on the winter solstice (December 20 or 21).

To resolve the lagging of quarter days, Khayyám devised an ingenious formula. His calendar interleaved eight extra days over each thirty-three-year span. The calculation: $365 + 8/33 = 365.2424$ days, aligned almost perfectly with the actual year length (365.2423 days), and proved even more accurate than that of the later Gregorian calendar: $365 + 1/4 - 1/100 + 1/400 = 365 + 97/400 = 365.2425$ days.

The Sultan officially adopted Khayyám's reform on Friday, March 15, 1079 (*Farvardin* 1, 458 according to the new calendar). Drums and cannon blasts across the nation proclaimed the first new year.

Concerning Khayyám's later years, we can say very little. The Sultan's premature death, just ten years after the calendar's adoption, brought an end to his generous patronage. Khayyám quit the royal court, only the ritual pilgrimage to Mecca delaying a return to his hometown. He continued to write poems.

> Ah, but my computations, People say,
> Reduced the year to better reckoning? Nay,
> 'Twas only striking from the Calendar
> Unborn tomorrow and dead Yesterday

One day in his eighth decade, Khayyám tired and lay down to rest. His heavy head swaddled in sheets of turban, he gazed up toward the sky. The light around him slowly faded, and then went out. The year was around 500 of the *hegira*—the point Al-Tabari had long ago calculated as being the end of the world.

Thirteen

COUNTING BY ELEVENS

Physicians say that thumbs are the master fingers of the hand," wrote Michel de Montaigne, the Renaissance nobleman who invented the personal essay, in his short piece *On Thumbs*. So vital did Rome's ancient rulers consider them, he explains, that war veterans who were missing thumbs received automatic exemption from all future military service.

Throughout his writings, the Frenchman marveled at the extent to which we depend upon our hands. A gesture of thumbs-up or -down, an index finger pressed to the mouth, the palms held flat open toward the skies: delivered at the right instant all can say more than any word. In another essay Montaigne describes the case of a "native of Nantes, born without arms" whose feet did double duty so well that the man "cuts anything, charges and discharges a pistol, threads a needle, sews, writes, takes off his hat, combs his head, plays at cards and dice." Elsewhere he observed that hands sometimes seem almost to possess a life of their own, as when his idle fingers would tap during a period of daydreaming, free of conscious effort or instruction.

Montaigne neglects to mention counting as being among the many useful tasks undertaken by our hands (biographers have pointed out that arithmetic was not his strong suit). Of course, there is much to be said for the idea that our decimal number system originated from the practice of counting with the fingers. In our Latin-derived word "digit," the meanings "number" and "finger" coincide. The Homeric Greek term for counting, *pempathai,* translates literally: "to count by fives."

People the world over, numbers at their fingertips, count to ten and by tens (twenty, thirty . . . fifty . . . one hundred . . .). But the manner of their getting to ten varies from hand to hand and from culture to culture. Like many Europeans, I count "one" beginning on the thumb of my left hand and continue past five with the digits of my right till I arrive at the other thumb. Where people read from right to left, as in the countries of the Middle East, counting generally begins on the little finger of the right hand. In Asia, the counter employs a single hand, folding the fingers thumb first to reach five, before unfolding them from the little finger (six) returning to the thumb (ten).

What, I wonder, would it be like for a person who had not too few fingers—like Montaigne's Roman veterans—but too many? Would such a person learn to count like you or me? How would it be to count by elevens?

According to tradition, Henry VIII's second wife, Anne Boleyn, carried a sixth finger on one of her hands—a medical condition known as polydactyly. Girls of high birth in Tudor society had the attention of tutors and learned to read, write, and do sums. Counting on her eleven fingers, however, would have posed Anne certain difficulties.

To start with, she would have found herself a number word short. Between the ninth finger and the last (which would still be the tenth—10 meaning "1 set and 0 remainders"), her surplus digit would have needed naming. Given that she spent part of her childhood in France, she might have come up with the label *dix*. So Anne would have counted: one, two, three, four, five, six, seven, eight, nine, *dix,* ten. Though the numbers' music sounds a little strange to our ears, in her mind the sequence would have come to feel like second nature. A year before the age of twenty, she would have celebrated (if only mentally) her dixteenth birthday. Counting higher, she would have taken the nineties down a peg, to make room for the dixties, and by way of dixty-dix arrive finally at one hundred.

Counting in this way produces some funny-looking sums. Subtracting seven from ten (where, given the extra *dix,* "ten" would to our way of thinking be eleven), for example, would have left Anne not with three, but four. Half of thirteen equals seven. Six squared (6 × 6) is thirty-three (three "tens" of eleven, plus three units): a pretty result.

Fractions would have proven particularly tricky. Unlike ten, eleven is a prime number: divisible only by itself and one. No precise midway point exists between one and eleven, or between Anne's ten (equivalent to our eleven) and her one hundred (11 × 11). What then might a half of something mean to someone with eleven fingers? And what about a fifth or a quarter or two-thirds? Possibly such concepts would remain as airy and intangible as a *dixth* of something might seem to you or me. (We can nonetheless imagine that rote learning would have sufficed to acquire them.)

Still, I am curious about the intuitions that an eleven-fingered girl might have brought to such ideas. From our two hands, equally endowed, we understand immediately that halves are clean and precise, leaving no remainder. Half of eight is four, not three or five. A right-angle triangle is exactly one half of a square. Prime numbers, by definition, cannot be halved in this way. Might Anne's hands—with six fingers on one hand, and "only" five on the other—have given her a more approximate, fuzzier feeling for the concept of a half? To a casual remark like, "I'll be along in half an hour," would the impatient thought have occurred to her, "Which half do you mean—the lesser or the greater?"

At her secret wedding to the King of England, Boleyn would have well understood which half of the couple she represented. Her triumph was infamously short-lived. Within months, the Archbishop of Canterbury had declared the marriage invalid. Roman Catholics denounced her as a scarlet woman.

The charge of treason brought against her only three years after the wedding makes no mention of witchcraft. All the same, her enemies alleged that the Queen's body showed strange warts and growths—the same body that had produced only stillborn fetuses for a male heir. It was in the role of a conspiring adulteress that she knelt in a black damask dress before the executioner's block on the morning of May 19, 1536.

It is certainly conceivable that the story of an eleventh finger was a fabrication of Anne's enemies. A famous portrait by an unknown artist, which now hangs in Hever Castle (Boleyn's childhood home) in Kent, reveals a striking young woman clutching a rose. The hands peek shyly from long sleeves; the fingers—ten in all—appear, in contrast to the smooth oval face, somewhat ill formed. A slander, or a secret, Anne's eleven fingers have become an integral part of her legend.

I was reminded of the legend recently when a fascinating newspaper article caught my attention. It was about one Yoandri Hernandez Garrido, a thirty-seven-year-old Cuban man with twelve fingers and twelve toes. Apparently, Fidel Castro's doctor once paid the extradexterous boy a visit and declared his hands and feet the most beautiful that he had ever seen.

In keeping with the Latin American taste for nicknames based on physical appearance, schoolmates gave Garrido the name "Veinticuatro" (twenty-four). He learned to count in class like his friends but one day, he recalls, his primary schoolteacher asked him for the answer to the sum 5 + 5. In his confusion he answered, "Twelve."

Garrido tells the reporter that he makes a fine living with his hands. American tourists regularly tender their dollars to have a photo taken with the twelve-fingered Cuban. A few extra bucks can go a long way in Castro's republic. Proudly, Garrido holds his hands up to the camera. There is generosity in his smile.

The article makes no mention of the adult Garrido's

arithmetic. Counting time, to give one example, should be simpler with twelve fingers—one digit for every hour on the clock. To find what time is nine hours before four o'clock in the afternoon, he need simply hold down five fingers (9–4) on his right hand to reveal the answer: one hand (six fingers) plus the right hand's thumb: seven a.m. One digit for every month as well: so half-year periods jump from one hand's finger to the same finger on the other hand.

We know that the Romans preferred to perform certain calculations using a base of twelve. In his *Ars Poetica* the poet Horace offers us a vignette of Roman boys learning their fractions.

> Suppose Albinus's son says: if one-twelfth is taken from five-twelfths, what is left? You might have answered by now: One-third. Well done. You will be able to manage your money. Now add a twelfth: what happens?
> One-half.

The Venerable Bede taught his fellow monks to quantify the various periods of time in the biblical stories using the Roman fractions. One-twelfth, he noted, went by the name of an *uncia* (from which comes the word "ounce"), while the remaining part of 11/12 was called *deunx*. Dividing into six, the sixth part (1/6) was called *sextans*, and *dextans* for the rest (5/6). *Quadrans* was the Roman name for a quarter (1/4); three-quarters they called *dodrans*.

What do we get if we add one *sextans* to a *dodrans* (1/6 + 3/4)? As quick as Horace's schoolboys, or the Venerable

Bede's monks, Garrido's fingers (or toes) would tell him: one-sixth equals two digits and three-quarters equals nine digits, so 1/6 + 3/4 = 11/12 (a *deunx*).

I wonder what Garrido makes of our deficient hands, all of us counting to the tune of ten? Does he pity us as the Romans pitied their thumbless veterans? He calls his condition a blessing; it is clear that he would not wish to be any other way.

Some believe that we should all learn to count like Garrido. A society formed in the mid-1900s advocates replacing the decimal system with a "dozenal" one, since twelve divides more easily than ten. Leaving aside itself and one, the number twelve can be divided into four factors: two, three, four, and six, compared with the factors of ten: two and five. In England and America the society still militates for the return of Roman fractions (among other measures), judging their abandonment a gross error.

Like the Esperantists and the simplified spellers that came before them, the society's members dream of a highly rational world comprising pairs, trios, quarters, and sextets, a world innocent of messy fractions. Theirs is the charm of the hopeless cause.

An English queen with eleven fingers, a Cuban man with twelve toes—such stories still incite our wonder and with it our vague sense that something has gone awry.

Montaigne, typically generous, remedies this view. He recalls once encountering a family who exhibited "a monstrous child" in return for strangers' coins. The infant had multiple arms and legs, the remains of what we now call a conjoined twin. In the infinite imagination of its Creator, Montaigne supposes, the child is simply one of a kind, "unknown to man." He concludes, "Whatever falls out contrary to custom we say is contrary to nature, but nothing, whatever it be, is contrary to her. Let, therefore, this universal and natural reason expel the error and astonishment that novelty brings along with it."

Fourteen

THE ADMIRABLE NUMBER PI

To believe the poet Wisława Szymborska, I am one in two thousand. The 1996 Nobel laureate offers this statistic in her poem "Some Like Poetry" to quantify the "some." Actually I think she is a tad too pessimistic — I am hardly as rare a reader as that. But I can see her point. Many people consider poetry to be all clouds and buttercups, without purchase on the real world. They are right and they are wrong. Clouds and buttercups exist in poetry, but they are there only because storms and flowers populate the real world too. Truth is, a good poem can be about anything.

Including numbers. Mathematics, several of Szymborska's verses show, lends itself to poetry. Both are economical with meaning; both can create entire worlds within the space of a few short lines. In "A Large Number," she laments feeling at a loss with numbers many zeroes long, while her "Contribution to Statistics" notes that "out of every hundred people, those who always know better: fifty-two" but also, "worthy of empathy: ninety-nine." And then there is "The Admirable Number Pi," my favorite poem. It — the poem, and the number — begins: three point one four one.

Once, in my teens, I confided my admiration for this number to a classmate. Ruxandra was her name. Like the poet's, her name came from behind the Iron Curtain. Her parents hailed from Bucharest. I knew nothing of Eastern Europe, but that did not matter: Ruxandra liked me. She liked that I was quite different from the other boys. We spent breaks between lessons in the school library, swapping ideas about the future and homework tips. Happily for me, her strongest subject was math.

In a moment of curiosity, I asked about her favorite number. Her reply was slow; she seemed not to understand my question. "Numbers are numbers," she said.

Was there no difference at all for her between the numbers 333, say, and fourteen? There was not.

And what about the number pi, I persisted—this almost magical number that we learned about in class. Did she not find it beautiful?

Beautiful? Her face shrank with incomprehension.

Ruxandra was the daughter of an engineer.

The engineer and the mathematician have a completely different understanding of the number pi. In the eyes of an engineer, pi is simply a value of measurement between three and four, albeit fiddlier than either of these whole numbers. In his calculations he will often bypass it completely, preferring a handy approximation like 22/7 or 355/113. Precision never demands of him anything beyond a third or fourth decimal place (3.141 or 3.1416, with rounding). Digits past the third or fourth decimal do not interest him; as far as he is concerned, it is as though they do not exist.

Mathematicians know the number pi differently, more intimately. What is pi to them? It is the length of a circle's round line (its circumference) divided by the straight length (its diameter) that splits the circle into perfect halves. It is an essential response to the question, "What is a circle?" But this response—when expressed in digits—is infinite: the number has no last digit, and therefore no next-to-last digit, no antepenultimate digit, no third-from-last digit, and so on. One could never write down all its digits, even on a piece of paper as big as the Milky Way. No fraction can properly express pi: every earthly calculation produces only deficient circles, pathetic ellipses, shoddy replicas of the ideal thing. The circle that pi describes is perfect, belonging exclusively to the realm of the imagination.

Moreover, mathematicians tell us, the digits in this number follow no periodic or predictable pattern: just when we might anticipate a six in the sequence, it continues instead with a two or zero or seven; after a series of consecutive nines, it can as easily remain long-winded with another nine (or two more nines or three) as switch erratically between the other digits. It exceeds our apprehension.

Circles, perfect circles, thus enumerated, consist of every possible run of digits. Somewhere in pi, perhaps trillions and trillions of digits deep, a hundred successive fives rub shoulders; elsewhere occur a thousand alternating zeroes and ones. Inconceivably far inside the random-looking morass of digits, having computed them for a time far longer than that which separates us from the big bang, the sequence 123456789...repeats 123,456,789 times in a row. If only we could venture far enough along, we would find the number's opening hundred, thousand, million, billion digits immaculately repeated, as though at any instant the whole vast array were to begin all over again. And yet, it never does. There is only one number pi, unrepeatable, indivisible.

Long after my schooldays ended, pi's beauty stayed with me. The digits had insinuated themselves into my mind. Those digits seemed to speak of endless possibility, illimitable adventure. At odd moments I would find myself murmuring them, a gentle reminder. Of course, I could not possess pi—the number, its beauty, or its immensity. Perhaps, in fact, it possessed me. One day, I began to see what this number, transformed by me, and I by it, could turn into. It was then that I decided to commit a multitude of its digits to heart.

This was easier than it sounds, since big things are often more unusual, more exciting to the attention, and hence more memorable than small ones. For example, a short word like *pen* or *song* is quickly read (or heard), and as quickly forgotten, whereas *hippopotamus* slows our eye (or ear) just enough to leave a deeper impression. Scenes and personalities from long novels return to me with far greater insistence and fidelity, I find, than those that originated in short tales. The same goes for numbers. A common number like thirty-one risks confusion with its common neighbors, like thirty and thirty-two, but not 31,415 whose scope invites curious, careful inspection. Lengthier, more intricate digit sequences yield patterns and rhythms. Not 31, or 314, or 3141, but 3 *1* 4 *1* 5 sings.

I should say that I have always had what others call "a good memory." By this, they mean that I can be dependably relied upon to recall telephone numbers, dates of birthdays and anniversaries, and the sorts of facts and figures that crowd books and television shows. To have such a memory is a blessing, I know, and has always stood me in good stead. Exams at school held no fear for me; the kinds of knowledge imparted by my teachers seemed especially amenable to my powers of recollection. Ask me for the third person subjunctive of the French verb "être," for instance, or better still the story of how Marie Antoinette lost her head, and I could tell you. Piece of cake.

Pi's digits henceforth became the object of my study. Printed out on crisp, letter-size sheets of paper, a thousand digits to a page, I gazed on them as a painter gazes on a favorite landscape. The painter's eye receives a near infinite number of light particles to interpret, which he sifts by intuitive meaning and personal taste. His brush begins in one part of the canvas, only to make a sudden dash to the other side. A mountain's outline slowly emerges with the tiny, patient accumulation of paint. In a similar fashion, I waited for each sequence in the digits to move me—for some attractive feature, or pleasing coincidence of "bright" (like 1 or 5) and "dark" (like 6 or 9) digits, for example, to catch my eye. Sometimes it would happen quickly, at other times I would have to plow thirty or forty digits deep to find some sense before working backward. From the hundreds, then thousands, of individual digits, precisely rendered and carefully weighed, a numerical landscape gradually emerged.

A painter exhibits his artwork. What was I to do? After three months of preparation, I took the number to a museum, the sprawling digits tucked inside my head. My aim: to set a European Record for the recitation of pi to the greatest number of digits.

March is the month of spring showers, and school holidays, and spick-and-span windows. It is also the month when people the world over celebrate "Pi Day," on March 14. So on that day, in 2004, I traveled north from London to the city of Oxford, where staff members at the university's Museum for the History of Science were waiting for me. Journalists, too. An article in the *Times,* complete with my photo, announced the upcoming recitation.

The museum lies in the city center, in the world's most ancient surviving purpose-built museum building, the Old Ashmolean. Iconic stone heads, wearing stone beards, peer down at visitors as they pass through the gates. The thick walls are the color of sand. Approaching the building, photographers appear as if from nowhere, holding cameras, like masks, up to their faces. The piercing flashes momentarily petrify my expression. I stop and raise my features into a smile. A minute later they are gone.

The record attempt's organizers have occupied the museum building. Television camera wires snake the length of the floor. Posters requesting donations (the event is raising money for an epilepsy charity, at my request, since I suffered from seizures as a young child) dress the walls. A table and chair, I see on entering, have already been set out for me on one side of the hall. Before it, a longer table awaits the mathematicians who will verify my accuracy. But there is still an hour before the recitation is due to start, and I find only a trio of men talking together. One has a full head of wiry hair, one has a multicolored tie, and one has neither hair nor tie. The third steps briskly forward and introduces himself as the main organizer, and the others as the museum's curator and his assistant. Their faces show mild puzzlement, curiosity, and nerves. Shortly afterward, reporters arrive to hold the microphones and man the television cameras. They film the display cases containing astrolabes, compasses, and mathematical manuscripts.

Someone asks about the blackboard that hangs high on the wall opposite us. The curator explains that on May 16, 1931, Albert Einstein used it during a lecture. What about the chalky equations? They show the physicist's calculations for the age of the universe, replies the curator. According to Einstein, the universe is about ten, or perhaps one hundred, thousand million years old.

Footfalls increase on the museum's stone steps as the hour approaches. The mathematicians duly arrive, seven strong, and take their seats. Men, women, and children keep coming; it is soon standing room only. The air in the hall grows thick with hushed talk.

At last, the organizer calls everyone to silence. All eyes are on me; nobody moves. I sip a mouthful of water and hear my voice begin. "Three point one four one five nine two six five three five eight nine seven nine three two three eight four . . ."

My audience is only the second or third generation able to hear the number pi beyond the first few tens or hundreds of decimal places. For millennia it existed only in a breathful of digits. Archimedes knew pi to only three correct places; Newton, almost twenty centuries later, managed sixteen. Only in 1949 did computer scientists discover pi's thousandth digit (following the decimal point): nine.

It takes about ten minutes, at a rate of one or two digits per second, for me to reach this nine. I do not know how long exactly; an electronic clock records the seconds, minutes, hours of the recitation for the public to watch, but I cannot see it from my chair. I stop reciting to sip water and catch my breath. The pause feels palpable. Dolorous even. I feel completely, oppressively alone.

The rules for the recitation are strict. I cannot step away from the desk, except to use the bathroom, and then always accompanied by a member of the museum's staff. No one may talk to me, not even to cheer me on. I can stop reciting momentarily to eat fruit or a piece of chocolate, or drink, but only at pre-agreed intervals a thousand digits apart. Cameras record my every sound and gesture.

"Three eight zero nine five two five seven two zero one zero six five four..."

An occasional cough or sneeze from the audience punctuates the flow of digits. I do not mind. I meditate on the colors and shapes and textures of my inner landscape. Calmness gains on me; my anxiety falls away.

Most of the spectators know nothing of Archimedes' polygons; they have no idea that the ten digits they have just heard will eventually repeat an infinite number of times, have never thought of themselves as being in any way susceptible to math. But they listen attentively. The concentration in my voice seems to communicate itself to them. Faces, young and old, round and oval, all wear delicate frowns. Listening to the digits, they hear their dress sizes, their birthdays, their computer passwords. They hear excerpts—both shorter and longer—from a friend's, or parent's, or lover's telephone number. Some lean forward in expectation. Patterns coalesce, and as quickly disperse, in their minds.

The people are all different. They have various motives for being here, and various goals. A teenager finds in the hall a hideout from his Sunday boredom; a manual laborer, having donated the equivalent of a pack of cigarettes in salary, sticks around to get his money's worth; an American tourist in shorts and a Mickey Mouse hat cannot wait to recount the spectacle to her family.

An hour passes, and then another.

"Zero, five, seven, seven, seven, seven, five, six, zero, six, eight, eight, eight, seven, six..."

I head further and further inside the number, exhaling effort, rhythm, and precision with every breath. The decimals exhibit a kind of deep order. Fives never outstrip sixes for long, nor do the eights and nines lord it over the ones and the twos. No digit predominates except for brief and intermittent instants. Every digit, in the end, has more or less equal representation. Every digit contributes equally to the whole.

Halfway through the recitation, more than ten thousand decimals in, I stop to stretch. I push back the chair, stand and shake out my limbs. The mathematicians put down their sharp pencils and wait. I bring a bottle to my lips and swallow the plastic bottle-tasting water. I eat a banana. I fold my legs, resume my position at the desk, and continue.

The silence in the hall is total. It reigns like a tsar. When a young woman's mobile telephone suddenly starts up, she finds herself promptly ejected.

Despite such rare commotions, a sly complicity establishes itself between the public and me. This complicity marks a vital shift. At the beginning, the men and women beamed confidence, listening expectantly, and taking pleasure in the digits' familiar sounds as those shoe sizes, historical dates, and car registration plates reached their ears. But, slowly, imperceptibly, something changed. Consternation grew. They could not follow the rhythm of my voice, they realized, without making continuous minor adjustments. Sometimes, for example, I recited the digits fast, and at other times I recited them slow. Occasionally, I recited in short bursts interleaved with pauses; at other moments, I recited the digits in a long, unbroken phrase. Sometimes the digits sounded thinly, accented by some inner agitation in my voice; instants later, they would soften to a clear and undulating beat.

Consternation now turns increasingly to curiosity. More and more, I feel the timing of their breathing coincide with mine. I sense their raw intrigue at the sound and sweep of every digit as it passes and makes way for the next. When the digits darken in my mouth—heavy eights and nines packed together—the tense distant faces grow tenser still. When a sudden three emerges from a series of zeroes and sevens, I hear something like a faint collective pant. Silent nods greet my accelerations; warm smiles welcome my slowdowns.

Between the moments when I stop reciting to sip water or take a bite, and continue reciting, I hardly know where to look. My solitude is absolute; I do not want to return the people's stares. I look down at the shadows of bones and veins in my hands, and at the scuffs in the wooden desk on which they rest. I notice the glimmers of shiny metal that dapple the display cases. On a cheek, here and there, I cannot help noticing, tears.

Perhaps the experience has taken the audience by surprise. No one has told them that they will find the number tangible, moving. Yet they succumb gladly to its flow.

I am not the first person to recite the number pi in public. I know there are a few "number artists"—men who recount numbers as actors recall their scripts. Japan is the center of this tiny community. In Japanese, spoken digits can sound like whole sentences; pronounced a certain way, the opening digits of pi, 3.14159265, mean "An obstetrician goes to a foreign country." The digits 4649 (which occur in pi after 1,158 decimal places) sound just like "nice to meet you," while a Japanese speaker pronouncing the digits 3923 (which occur after 14,194 decimal places) simultaneously says, "Thank you, brother."

Of course, such verbal constructions always suffer from arbitrariness. The short, stiff phrases stand apart, with only the speaker's ingenuity to splice them together. Japanese spectators, I have heard, watch these men perform as they might watch a tightrope walker; listening only in case of a blunder, as others watch only in case of a fall.

The relationship these artists have with numbers is complicated. Many years of repetitive learning hone their technique; but they also produce a nagging feeling of duplicity: repeated numbers (and words) often finish by losing all their sense. It is not uncommon, after each public display, for the performer to impose a monthslong fast of every digit. Benumbed by numbers, even a price tag, a barcode, an address, sickens him.

In the number artist's brain, pi can be reduced to a series of phrases. In my mind, it is I, not the number, who grows small. I diminish myself as much as possible before the mystery of pi. Emptying myself, I perceive every digit up close. I do not wish to fragment the number; I am not interested in breaking it up. I am interested in the dialogue between its digits; in the unity and continuity that underlie them all.

A bell cannot tell time, but it can be moved in just such a way as to say twelve o'clock—similarly, a man cannot calculate infinite numbers, but he can be moved in just such a way as to say pi.

"Three, one, two, one, two, three, two, two, three, three, one..."

Reciting, I try to summon up a true picture of what I see and feel. I want to convey the shapes, and colors, and emotions that I experience, to everyone in the hall. I share my solitude with those who watch and listen to me. There is intimacy in my words.

A third hour comes and goes; the recitation enters its fourth hour.

More than sixteen thousand decimal places have escaped

my lips. Their swelling company presses me on. But exhaustion also grows within my body, and all of a sudden my mind goes blank. I feel the blood falling out of my head. Up until only a few moments ago the digits had accompanied me; now they make themselves scarce.

In my mind's eye, ten identical-looking paths stretch out before me; each path leading on to ten more. One hundred, a thousand, ten thousand, one hundred thousand, a million paths, beckon me out of the impasse. They stream in every possible direction. Which way to go? I have no idea.

But I do not panic. What good did panicking ever do anyone? I shut my eyes tight, and coaxingly rub the skin around my temples. I take a deep breath.

Green-tinted blackness pervades my mind. I feel disoriented, lost. A filmy white surfaces over the black, only to be recovered by a rolling gray-purple. The colors bulge and vibrate but resemble nothing.

How long did these maddening misty colors last? Seconds, but they each seemed agonizingly longer.

The seconds pass indifferently; I have no choice but to endure them. If I lose my cool, all is finished. If I call out, the clock comes to a halt. If I do not give the next digit in the next few moments, my time will be up.

No wonder the next digit, when finally I release it, tastes even sweeter than the rest. This digit requires all my force and all my faith to extract it. The mist in my head lifts, and I open my eyes. I can see again.

The digits flow fleet and sure, and I regain my composure. I wonder if anyone in the hall noticed a thing.

"Nine, nine, nine, nine, two, one, two, eight, five, nine, nine, nine, nine, nine, three, nine, nine..."

Quickly, quickly, I must keep going. I must not let up. I cannot linger, not even before the most outstanding glimpses of the number's beauty; the joy I feel is subordinate to the need to reach my goal and recite the final digit in my mind. I must not disappoint all who are standing here, watching me and listening to me, waiting for me to bring the recitation to its fitting conclusion. All the preceding thousands have no value in themselves: only once I have wrapped everything up can they successfully count.

Five hours have now elapsed. My speech begins to slur; I am drunk on exhaustion. The end is in sight, but it generates fear: am I up to it? What if I fall short? Tension stirs me for this culminating burst.

And then, minutes later, I say, "Six, seven, six, five, seven, four, eight, six, nine, five, three, five, eight, seven," and it is over. There is nothing more to say. I have finished recounting my solitude. It is enough.

Palms come together; hands clap. Someone lets out a cheer. "A new record," someone else says: 22,514 decimal places. "Congratulations."

I take a bow.

For five hours and nine minutes, eternity visited a museum in Oxford.

Fifteen

EINSTEIN'S EQUATIONS

Speaking about his father, Hans Albert Einstein once said, "He had a character more like that of an artist than of a scientist as we usually think of them. For instance, the highest praise for a good theory or a good piece of work was not that it was correct nor that it was exact but that it was beautiful." Numerous other acquaintances also remarked on Einstein's belief in the primacy of the aesthetic, including the physicist Hermann Bondi, who once showed him some of his work in unified field theory. "Oh, how ugly," Einstein replied.

It is a mostly thankless task to try to assign to mathematicians some universal trait. Einstein's famous predilection for beauty offers one rare exception. Mathematicians can be tall or short, worldly or remote, bookworms or book burners, multilingual or monosyllabic, tone-deaf or musically gifted, devout or irreligious, hermit or activist, but virtually all would agree with the Hungarian mathematician Paul Erdös when he said, "I know numbers are beautiful. If they are not beautiful, nothing is."

Einstein was a physicist, yet his equations inspired the

interest and admiration of many mathematicians. His theory of relativity drew their praise for combining great elegance with economy. In a handful of succinct formulas, every symbol and every number obtaining its perfect weight, Newtonian time and space were recast.

Books on popular mathematics abound with discursive explanations of technical proofs to illustrate their beauty. I cannot help but wonder if this might not be a mistake. I suspect that, more usually, what we laymen really admire in the work of a Euclid or an Einstein is its ingenuity, rather than its beauty. We are impressed, and yet unmoved by them.

The barrier to an appreciation of mathematical beauty is not insurmountable, however. I would like to suggest a more indirect approach. At a remove from the technical acumen of a theorist, my suggestion is more intuitive. The beauty adored by mathematicians can be pursued through the everyday: games, and music, and magic.

Take the game of cricket, which was the frequent inspiration for G. H. Hardy—a major number theorist and the author of *A Mathematician's Apology*—who scoured the newspaper for cricket scores over breakfast every morning. In the afternoon, after some hours at his desk, he would furl his theorems and transport them in his pockets (in case of rain) to see a local match. Among his papers he sketched the following cricket "dream team":

Hobbs

Archimedes

Shakespeare

M Angelo

Napoleon (Capt)

H Ford

Plato

Beethoven

Johnson (Jack)

Christ (J)

Cleopatra

Cricket matches presented Hardy the spectator with the same "useless beauty" that he so cherished in his theorems. By useless, he meant only that neither had any goal beyond the pursuit for its own sake. He would also frequently stand at the stumps himself, surrounded by the other team's fielders, watching the red ball expand as it flew toward his bat. Both experiences seemed to stimulate his mathematical antennae for order, pattern, and proportion.

Time elapses differently on a cricket ground or in a concert hall. At its best, a well-executed, smooth-flowing cricket match can replicate the sense of harmony that we most often associate with music. The tension mounts and falls tidally, like the notes in a song. A five-day match is adept at slackening and pulling tight the outline of its hours, while every musical composition bears its own time within the structure of its notes. The unique tempo is also a part of the experience of mathematical beauty.

Gottfried Leibniz wrote that music's pleasure consisted of "unconscious counting" or an "arithmetical exercise of which we are unaware." The great philosopher-mathematician meant, I suppose, that the numerical ratios that underlie all music are grasped intuitively by our minds. Every instant the listener mentally resolves the relation between the various notes—the fourths and fifths and octaves—as though they were objects all laid out before him side by side in some gigantic illustration. This "grasping" of the music—however fleeting—is something we can all experience as beautiful.

On the relationship between musical and mathematical beauty, we can learn more from writings about Pythagoras. It is said that he possessed a musician's ear. From boyhood, he showed a flair for the lyre. Perhaps he first heard its seven strings played by a traveling *citharede,* a female performer dressed in long curls and bright colors; the *citharede*s were the divas of their day.

Pythagoras discovered that the most harmonious notes result from the ratios of whole numbers. A vibrating string exactly halved or doubled, for example, produces an octave (ratio 1:2 or 2:1). If precisely one-third of the string is held down, or when the string is tripled in length, a perfect fifth (an octave higher) results. A perfect fourth can be obtained by holding down one-quarter of the string, or stretching it out four times longer. The whole harmonic scale was constructed in this way. Pythagoras observed that all of music depended on the first four numbers and their interrelations. He worshiped ten as the most perfect number, reflecting the unity of all things, it being the sum of one and two and three and four.

According to Hippolytus, one of the great theologians of the Early Church, Pythagoras taught that the cosmos sang and was composed of music and "he was the first to put down the movement of the seven stars to rhythm and melody." He even attempted to reproduce this soothing universal music for his disciples, rousing them in the morning to the sound of his lyre. In the evening, too, he would play for their benefit, "in case too turbulent thoughts might still inhabit them."

From Pythagoras's lyre it is an easy hop to Einstein's violin. "If I were not a physicist," he once told an interviewer, "I would probably be a musician. I live my daydreams in music. I see my life in terms of music. I get most of my joy in life out of music." His violin case accompanied him on many trips, but Einstein played with discretion and few reliable reports of his musical ability survive. It appears that his amateur technique was respectable, if somewhat limited. Though there may be an affinity between the laws of math and music, they cannot be conflated. Even prodigious mathematical gifts such as Einstein's did not translate into exceptional musicianship, but they certainly sharpened and reinforced his appreciation.

If math is the secret underpinning the harmonies of cricket and music, mathematical beauty is also key to feats of magic. Ever since I was a small child, playing cards that swap places, white handkerchiefs that take wing, and top hats that turn into rabbit holes have fascinated me. The visions touch some nerve deep inside.

One evening several years ago I attended the London performance of a young conjuror, who was playing to a full house. I was seated in one of the middle aisles, in a sea of heads, next to a rather elderly gentleman whose stomach sat on his lap. From this distance I had a good view of the stage. A blend of high technology spotlights and music hall gloom favored illusion.

People come to magic shows for all sorts of reasons: some for the theatrics, others for the performer's comedy, and some (like my neighbor), it would seem, to cough. The main draw for me, though, is an experience of the unexpected. It is this that gives a conjuror's performance its peculiar beauty, a beauty akin to a well-turned equation.

I am not talking here about the end result, the "effect," of a trick. I am talking about the method. Every mind-read drawing or woman sawn in half looks pretty much alike, whereas the hidden ideas that make each possible can vary as much as their performers. Dozens, if not hundreds, of ways exist to levitate a spoon or make the Statue of Liberty vanish, in the same manner that numerous hundred people (not all of them professional mathematicians) have shown that the square of a right-angled triangle's hypotenuse equals the sum of the squares of the other two sides. But few of these theoretical demonstrations, or magical methods, would be considered up to Einstein's test of beauty. The truly beautiful are those that foster surprise.

A genuine experience of the unexpected, in math as

much as in magic, demands of its performer at once originality of insight and a lightness of touch. Even a single step too many in a method renders ugly and clumsy the theorem or the trick.

It is sometimes said that a magician will go to great pains to conceal his workings from the public. The reality is that only a poor method requires such attention; a fine one, by its beauty, conceals itself. We might call this rule the coquetry of good technique.

One part of the magic performance from that evening in London well illustrates this point. A shy-looking woman from the audience was invited up onto the stage. At its center, on a pedestal, stood a glass bowl containing large multicolored buttons. The buttons, we were told, totaled one hundred in all. Following instructions, the woman dipped her hands into the bowl, netting as many of the buttons as she wished. Unclenching her fingers, she next poured the buttons onto a tray and covered them with a tea towel. The magician approached the tray, peeked under the towel for all of two seconds, and then turned to the public.

"Seventy-four," he declaimed.

The woman duly proceeded to count the scattered buttons on the tray, one by one. It took quite some time. After a minute or so, she prodded the final button and the expression of her face lengthened. There were precisely seventy-four buttons on the tray. Gasps popped around my ears, followed immediately by a brisk weather front of applause. The "counting buttons" trick was a highlight of the show.

I imagine some people applauded the magician for mental powers that flirted with the supernatural. To recognize seventy-four items (and not mistake them for seventy-three, or seventy-five) in the space of two seconds is, it must be said, quite something. Neurologists tell us that the human brain "subitizes" (counts at a glance) no better than in fours or fives. This figure remains constant across all categories of people, irrespective of training or synaptic quirks—neither mathematicians nor autistic savants exceed it. In two seconds, even the most practiced eye can enumerate only eight or ten, and no higher.

The explanation of divination did not cross my mind; even admitting that it was possible (which I do not), it would have felt somewhat disappointing. A laborious counting of every button, even if accomplished at the most remarkable clip, would be totally lacking in finesse or beauty. I groped with my imagination for the magician's insight.

How might a person count some (relatively) large quantity in no time at all? This question stayed with me that night as I tossed and turned in bed, finally to sleep. In my dreams the gleaming transparent globe and its outsize buttons and the shy woman holding the tray all returned to me. I looked and looked but failed to see.

Early in the morning I woke to a beautiful sensation of absolute clarity. The night had seemingly done its work. Now every moment of the magician's trick, from beginning to end, made perfect sense. Had I unveiled the trick's machinery? I could not say for sure. In its simplicity and economy the solution felt inevitable, but I have no idea whether it was in fact the magician's or merely my own. In any event my mood that morning was immediately elevated. I wanted, like Archimedes in his bath, to leap and cry, “Eureka!” Quite possibly I did. I felt like a mathematician who receives the sudden, shocking ecstasy of a proof.

Of the school of nocturnal thoughts that swam anonymously through my brain, one image had stuck. It was a modest household object that I used every day: my kitchen scale. Now to the question “how might a person count some large quantity in no time at all?” came at once the delicious answer: weigh it! What if the identical buttons weighed exactly one gram each? And what if the glass bowl’s pedestal concealed a weighing scale? When the woman lifted seventy-four buttons from the bowl, the reading (displayed backstage) would fall instantly from “100” to “26.” This number, via an earpiece or some prearranged signal, could then be relayed to the performer. That the math involved might be the simplest of simple subtractions only accentuates the solution’s charm.

This pure beauty that we call mathematical, and that we find in games, music, and magic tricks, is something like a rumor or a longing that lingers in the person, hinting at significance and depth. We go back to it time and time over: it is beautiful, because it stays. It is we who change.

Problems, in magic or mathematics, are wonderful things. Without problems, we would have no proofs, and the shimmering pleasure of elucidation is a thing of beauty. Einstein’s equations possessed this special quality in abundance, of course. $E = mc^2$ (energy is equal to mass times the speed of light squared) answered riddles—such as the behavior of light — that most other scientists had not even seen.

I have spoken about mathematical beauty with hardly any reference to numbers, but, of course, numerical problems also provide many instances of beauty. A personal example, from arithmetic, is the multiplication 473 × 911. The solution — 430,903 — may at first sight appear merely banal. Its repeating threes and zeroes, however, reversed as in a mirror, hint at some attractive pattern lurking beneath. And so we begin to loiter around the answer. With a closer look, we might observe the relationship: 903 – 430 = 473. If now we modify the original question slightly, 473 × 910, simplifying the sum, we reach the answer: 430,430. And we ask ourselves: how is this possible? Again we return to the original problem and dissect its numbers. 473 equals 43 × 11. The number 910 comprises 7 × 13 × 10. We toy with these smaller constituents, until at last we discover that our original sum equates to (430 × 1001) + (43 × 11).

A further illustration of this numerical beauty can be found within the primes. The number 75,007 (as it happens, a rather chic Parisian postcode) poses the problem of whether any smaller number evenly divides it. In other words, is the number 75,007 prime? Deceptively simple to formulate, this question proves to be treacherously difficult to answer. As with the sum above, we must grapple with the number until its secret finally yields.

We begin by assuming (the odds are on our side) that the number 75,007 is composite — that some smaller numbers exist that will cleanly divide it. Not an even number, it cannot be divisible by 2 (the smallest prime). The number 75, we note, divides into 3 and 5 (the next primes), but not when followed by two zeroes and a seven. We might think of 75,007 as a house on a very lengthy street, and observe that, sixty-eight doors down, number 75,075 patently divides by 1,001 (and therefore, by 7 and 11 and 13). But 68 can be split only into two and then a further two to leave seventeen.

Imagine the mathematician, pencil in hand, attempting to winkle out the number's factors. The insight resists him; he wonders whether it will ever come. He abandons his desk, pacing the floor, and he marches along the mental street and its numbered houses. And then, all of a sudden, the dizzying thought strikes him: 75,007 can be expressed as 74,900 + 107, or (10,700 × 7) + 107, or more precisely still (107 × 100 × 7) + 107. His blood leaps with joy to recognize the repeated factor: 107. On a crisp sheet of paper, he writes: 75,007 = 107 × 701.

Human beings' quest for meaning is perpetual; lack of meaning is offensive to the mind, and whatever the scale of the problem, a solution is a thing of beauty. Einstein's equations solved problems such as "What do we mean by the words 'time' and 'mass'?" A mathematician could tell us that the number 75,007 means to travel from 0 to 107, and then repeat the same distance successively 701 times. Other meanings, like those found in music or cricket, while more intimate and inexpressible, can prove just as powerful. Where chaos is subdued and the arbitrary averted, there lies beauty, and it is all around us.

I recall an afternoon I once spent at a summerhouse belonging to friends. We had just returned, tired and hungry, from a long hike among the encircling hills. One of my friends switched on a little radio. We sat here and there around the living room, with views out to the sea, half listening and half talking. The broadcaster was reading out the week's letters sent in by listeners to the program. Among their litany of compliments and complaints, he recited a short puzzle sent in by a longtime listener in the north.

"A remarkable species of water lily doubles in size every day. If the lily covers an entire lake in a period of 30 days, after how many days will it cover half of the lake?"

Our various discussions slowed, then continued as before. Someone switched off the radio. The face of the friend opposite me gradually darkened, and her replies became wayward and staccato. Other voices took up the slack. It seemed no one else had taken the problem of the lily to heart.

Minutes passed. My friend cast sidelong glances at the walls and the windows and the flower-lit hills they framed. Her blue eyes narrowed to a squint. Kitchen noises filled the house, followed by the chattering distribution of hot teacups, exhaling steam. Her untouched drink trembled when she struck the table absentmindedly with her shin.

And then I saw it. Suddenly, her features shone. "Twenty-nine," she said, with a broad smile. If the lily doubles every day, it will also halve every day when we track its history. With a size of 1 (for "one lake") after 30 days, it had a size of

0.5 after 29 days, 0.25 after 28 days, 0.125 after 27 days, and so on.

The moment had surprised her, she said. It had come right out of the blue. In this moment, I saw my friend behold the astonishing beauty of mathematics.

Sixteen

A NOVELIST'S CALCULUS

The history of the world, Tolstoy said, is the history of little people. Leo Nikolayevich himself, though, was a giant of a man. Just shy of six feet, he was taller than most of his contemporaries. Stronger, too. He could lift 180 pounds with a single hand. He dressed his muscles plainly, in a peasant's smock, with a belt that cinched the small of his back. His ego-enhancing arguments were just as sturdy. Always resistant to the thinking of his day, he denounced historians as hero worshipers. In *War and Peace,* over one thousand pages long, he launched his most sustained attack. His primary weapon was drawn from mathematics.

Calculus was by no means a novel idea in Tolstoy's time. Its "inventors," Isaac Newton and Gottfried Leibniz, in the late seventeenth century, were refining theories that had been in development since the time of the ancient Greeks. As geometers study shape, the student of calculus examines change: the mathematics of how an object transforms from one state into another, as when describing the motion of a ball or bullet through space, is rendered pictorial in its graphs' curves. In these curves, smooth and subtle, girding the infinitesimal movements behind every human life, Tolstoy thought he saw the blindness of contemporary historians.

Alongside formidable intellectual powers, he certainly had a head for exotic ideas. I think of the wackier pronouncements: the dismissal of Shakespeare as a lousy poet, of Darwinism as a transient fad, of marriage as legalized fornication. Like Thomas Jefferson, he took a pair of scissors to the New Testament, trimming its pages of every one of its miracles. His cult of simplicity (as G. K. Chesterton later called it) welcomed flocks of disciples to his estate: men and women, young and old, dressed in bedsheets and bark sandals, who followed his every step and hung on his every word. But more ambitious, more inventive, and more subversive than any of these was the novelist's view of history itself as a kind of calculus.

We find this view studded throughout the pages of *War and Peace,* in those portions that resemble the tight and intense arguments of a pamphleteer. It so happens that they are the same parts that most modern readers tend, perhaps understandably, to skip. But this less diligent reader misses a crucial bedrock of Tolstoy's work.

> The movement of humanity, arising as it does from innumerable arbitrary human wills, is continuous. To understand the laws of this continuous movement is the aim of history... only by taking infinitesimally small units for observation... and attaining to the art of integrating them (that is, finding the sum of these

infinitesimals) can we hope to arrive at the laws of history.

Calculus, which Tolstoy defined as "a modern branch of mathematics having achieved the art of dealing with the infinitely small," offered him a vocabulary in which to voice his disagreement with many historians. He denounced their lamentable tendency to simplify. The experts stumble onto a battlefield, into a parliament or a public square, and demand, "Where is he? Where is he?" "Where is who?" "The hero, of course! The leader, the creator, the great man!" And having found him, they promptly ignore all his peers and troops and advisers. They close their eyes and abstract their Napoleon from the mud and the smoke and the masses on either side, and marvel at how such a figure could possibly have prevailed in so many battles and commanded the destiny of an entire continent. "There was an eye to see in this man," wrote Thomas Carlyle about Napoleon in 1840, "a soul to dare and do. He rose naturally to be the King. All men saw that he was such."

But Tolstoy saw differently. "Kings are the slaves of history," he declared. "The unconscious swarmlike life of mankind uses every moment of a king's life as an instrument for its purposes." Kings and commanders and presidents did not interest Tolstoy. History, his history, looks elsewhere: it is the study of infinitely incremental, imperceptible change from one state of being (peace) to another (war).

The experts claimed that the decisions of exceptional men could explain all of history's great events. For the novelist, this belief was evidence of their failure to grasp the reality of an incremental change brought about by the multitude's infinitely small actions. Out of a need to theorize, to locate "causes," the historian privileges one series of events and examines it apart from all the others. Why, all of a sudden, had Napoleonic France and Tsarist Russia rushed to war? What drove millions of men, men who licked their plates and read stories to their sons and worried about their looks, to suddenly thieve and crush and slaughter one another? Napoleon overreached, a victim of his own pride and mania, says one expert. He let himself go, growing fat and moody. With successive battle victories under his belt, he fell inevitably to thinking himself invincible. No, no, says another historian, you forget how weak and highstrung was the Tsar Alexander. Such weakness certainly invited a military strike. The long-standing economic embargoes in Europe, suggests a third, led to strained relations between the different peoples. A fourth points out that hundreds of thousands of soldiers obtained gainful employment. Napoleon himself, near the end of his life, is said to have put the war down to the intrigues of the British.

Naturally, not all these "causes" can be right, and some are even mutually contradictory. Either Napoleon's decision to invade Russia was impetuous and instinctive, or else it was carefully calculated (against Russian weakness) and deliberate (to keep his forces busy). Either Russian weakness impelled the attention of France's army, or else Napoleon's mania invented such weakness for its own purpose. Either the war resulted from French initiative, or from British interference.

As a Briton who lives in France, I see how each nation selects its own causes, and edits them convincingly into its own version of history. In Britain, Napoleon's name is synonymous with tyranny and a small comic man's delusions of grandeur. In France, *au contraire,* he is a revolutionary who stood up for the new Republic against the hostile monarchies of Europe. The puffed-up Napoleon with "small white hands," as depicted by Tolstoy, is, of course, a Napoleon from the Russian perspective.

This third Napoleon, as conceived by Tolstoy, had at least one cardinal virtue: he knew to keep out of the way of his soldiers, not to tread on anyone's boots, to give a fair impersonation of someone who is in command. It is the soldiers who shoot, and stab, and cough, and groan and bleed. They constituted the vast majority of the French emperor's army, but they issued not a single command. The commands came from the officers above the soldiers, who took them in turn from the generals above the officers, who took them from the commander-in-chief above the generals. The most important commands always come from those who participate least in the physical action. Consequently, most of these commands, thousands of them, not corresponding to conditions "on the ground" at the moment and in the place that they finally filtered down to the troops, were never executed. They did not coincide with the reality of circumstances that remained beyond the chief's control. As far as Tolstoy is concerned, then, to say that Napoleon invaded Russia is only to say that a few of his commands, out of the thousands that came to nothing, coincided with certain broader events between the peoples of France and Russia in the year of 1812.

What were these broader events "on the ground"? As the novel's calculus analogy suggests, they were innumerable, infinitesimal. At a given moment, in a given place, the wishes and desires and intentions of hundreds and thousands of people temporarily coalesced. Tolstoy illustrates such a moment in the life of a backwoods Russian region.

> In the vicinity of Bogucharovo were large villages belonging to the crown or to owners whose serfs . . . could work where they pleased . . . In the lives of the peasantry of those parts the mysterious undercurrents in the life of the Russian people, the causes and meaning of which are so baffling to contemporaries, were more clearly and strongly noticeable than among others. One instance, which had occurred some twenty years before, was a movement among the peasants to emigrate to some unknown "warm rivers." Hundreds of peasants . . . suddenly began selling their cattle and moving in whole families toward the southeast. As birds migrate to somewhere beyond the sea, so these men with their wives and children streamed to the southeast, to parts where none of them had ever been. They set off in caravans, bought their freedom one by one or ran away, and drove or walked toward the "warm rivers." Many of them were punished, some sent to Siberia, many died of cold and hunger on the road, many returned of their own accord, and the

> movement died down of itself just as it had sprung up, without apparent reason. But such undercurrents still existed among the people and gathered new forces ready to manifest themselves just as strangely, unexpectedly, and at the same time simply, naturally, and forcibly. Now in 1812, to anyone living in close touch with these people it was apparent that these undercurrents were acting strongly and nearing an eruption.

Contemporary historians, to believe Tolstoy, took no notice of these "undercurrents" in the life of a people. They failed to see the ocean of history for the waves. Aware only of the tides they called "causes," they ignored the vast depths from which these ripples emerge. A man called Napoleon has an impetuous character; six months later Moscow is under siege. The historian looks at these two situations and asserts a link: namely, that hundreds of thousands of Muscovites fled their homes, and whole battalions of soldiers lost their lives, because of the impetuousness of this single man called Napoleon. Or, the historian notices that in, say, Liverpool and London there occurred local riots caused by a shortage of bread and that within a year Russian troop masses were fending off the French. Entire theories, each more elaborate and ingenious than the next, are spun in order to thread the rowdy fisticuffs in these English cities with the subsequent slashing and burning and murdering at Borodino.

I have, admittedly, offered only gross approximations of these historical theories, and of course they are often far more complex, discerning many separate causes—one cause after another—of a war. The temperament of a man called Napoleon is only one cause, they say, and the bread shortage in a city like Liverpool another. Often they will find a third, and perhaps a fourth or fifth, cause to supplement the first and second. All the same, Tolstoy's chief objection remains. Historians tend by their very nature to adopt a flawed approach, he argued, because a mass conflict can no more be reduced to a handful of causes than can a ship's course be reduced to a few waves. Between a French port and a Russian port lie innumerable points in the sea: why label the fifteen thousand four hundred and third point, say, or the seventy-one thousand nine hundred and sixty-eighth point, as being ultimately responsible for the ship's arrival?

An equivalent mistake would be to inquire of an age-beaten man, in which hour of your life was the blow delivered? Which blow? Why, the blow that loosened your teeth, and broke your bones, and thrashed your skin. Of course, there is no sense in such a question. The flow of time erodes patiently, continuously. What, then, could our elderly man say by way of a reply? He might recall that during a particularly hot summer night in 1968 he rolled out of bed and broke a shin. Perhaps he would smell once more the harsh carbolic soap with which, as a child growing up in the 1940s, he scrubbed his face. A game with his grandson in 1997 might come to his mind, in which a hard rubber ball accidentally struck his jaw. But none of these events, not less in combination, could truly help us to understand the elderly man's present condition.

Change appears to us mysterious because it is invisible. It is impossible to see a tree grow tall or a man grow old, except with the precarious imagination of hindsight. A tree is small, and later it is tall. A man is young, and later he is old. A people are at peace, and later they are at war. In each case, the intermediate states are at once infinitely many and infinitely complex, which is why they exceed our finite perceptions.

Even a dramatic change can thus be accomplished without our knowledge. A friend once related to me the following illustrative tale. A friend of my friend inherited a house in southern Europe. This house contained many fine pieces of furniture and works of art. Every summer, she flew to Europe and lived for several months in the house among these objects. She sat on the same cushions, walked past the same paintings, and heard the same ticking of the grandfather clock. The house's upkeep she entrusted to a small and loyal staff, so that its rooms were always cleaned and polished and in good repair whenever she came through the door. And then one day, several summers following her inheritance, her kid sister came to stay. The sister felt excited; she had heard many good things about the house and was longing to see it. But this feeling quickly gave way first to curiosity, then to confusion, and finally to astonishment. A distinguished-looking chair in the hall, upon closer inspection, revealed itself to be cheap and rickety. Removed from its frame, the painting that hung above the fireplace flapped poster-thin. The marble-colored statuette in the guest bedroom gave off the unmistakable whiff of plastic. Fakes! Frantically, the sisters raced from room to room, until the whole place was left upside down. Every chair, every vase, every painting, virtually everything in the house—over one hundred items—unbeknownst to this woman, had been meticulously replaced. Little by little, piece by piece, a wily member of the staff had stolen the house away from under her nose.

Sometimes, revolutions turn over a country in the same way that this staff member turned over the house. Imperceptibly, dissidence grows across a land long before the dictator calls out his tanks. And as recent events in the Arab world remind us, nobody predicts a revolution before it happens and nobody controls it once it is under way. "Why war and revolution occur we do not know," affirms Tolstoy. "We only know that to produce the one or the other action, people combine in a certain formation in which they all take part." Hearing the sudden drumbeat of shoes, the roar of voices, the upturned faces flushed with anger, the almighty, most wise, and beneficent ruler does not comprehend. In his incomprehension, he asks the same question that many of his fellow dictators end up asking themselves: from where did they come? All these people with their fists and their voices. He shakes his head in disbelief. And yet, the answer is simple because only one answer is possible. Simply put, the people were always there: on the streets, in the mosques, in the bazaars. Only now, this distributed mass of people suddenly and noisily combines. Only now do the man without work, and the woman without dignity, and the teenager without anything to eat make themselves seen and heard.

What is the power that moves peoples? Not the power of rulers, says Tolstoy, or for that matter the power of ideas. It is ineffable, invisible.

> History seems to assume that this force is self-evident and known to everyone. But in spite of every desire to regard it as known, anyone reading many historical works cannot help doubting whether this force, so variously understood by the historians themselves, is really quite well known to everybody.

This force is the Human Life in which every person—from the lowly peasant Karataev to the emperor Napoleon—participates. It is the "hidden warmth" of patriotism experienced by the Muscovites when suddenly confronted with the dramatic threat of foreign invasion. It is the "chemical decomposition" of the fleeing mass of French soldiers once their goal (the head-on confrontation in Moscow with an appointed "enemy") becomes unattainable. It is the "force of habit," during a salon conversation, which makes Prince Vasili Kuragin say "things he did not even wish to be believed."

Rather than assign various degrees of responsibility to this cause or that, Tolstoy proposes that historians pay far greater attention to this force. Moscow's conflagration, which the historians variously explain as a defensive tactic of the Russians (the so-called scorched-earth policy) or the wild vengeance of the invading French, becomes explicable in other terms.

> Moscow was burned because it found itself in a position in which any town built of wood was bound to burn.... A town built of wood, where scarcely a day passes without conflagrations when the house owners are in residence and a police force is present, cannot help burning when its inhabitants have left it and it is occupied by soldiers who smoke pipes, make campfires of the Senate chairs in the Senate Square, and cook themselves meals twice a day.

Readers were shocked by these arguments when the book first appeared (over several editions) in the 1860s. Turgenev denounced the historical reflections as "charlatanism" and "puppet comedy," while Flaubert found them merely repetitive. The historian A. S. Norov entitled his review of the book "Tolstoy's Falsification of History." Another historian, Kareev, complained that the novelist wanted to abolish history altogether. When, one and a half centuries later in the year 2000, a Russian publisher edited and released an early draft of the book, it boasted about the lack of all these troublesome elements. "Half the usual length. Less war and more peace. No philosophical digressions or incomprehensible French. A happy ending."

Mathematicians, on the other hand, have proved considerably more receptive. Tolstoy's mathematician friend, Urusov, expressed his delight at the analogy with calculus.

And more recently, in a 2005 article for the Mathematical Association of America, Stephen T. Ahearn praised Tolstoy's mathematical metaphors as being both "rich" and "deep" and encouraged math teachers to use them in their classrooms.

What, then, might we conclude? Should we even conclude? After all, if Tolstoy is right, his book—like any event in time—cannot be understood with prior assumptions, rules, and theories. Everything has its moment, its context. Earlier, in one state, you began this essay, and now later on, you finish it in another. What do you think? I cannot tell you. In everyone and everything, the process of change always asserts its own meaning.

Seventeen

BOOK OF BOOKS

I have walked in my sleep, and talked in my sleep, but I have never written in my sleep. The Icelandic author Gyrðir Elíasson's short story "*Næturskriftir*" ("Nightwriting") depicts a character whose writer's block suddenly disappears once the lights go out. In a notebook lifted from his bedside table, he begins to write down words, sentences, and even whole stories while he dreams.

> The days passed by all the same; he could not write . . . but at night he wrote; nearly every night . . . his wife knew not to wake a sleepwriter, so she lay [in the bed] and watched the expression of his back, how he wrote with amazing confidence with the notebook on his knees. (My translation).

Something about Elíasson's tale touches a chord in me. I think it has to do with confronting the infinity that is every written and unwritten book, including the "Book of Life": the infinitely many potential combinations that comprise our days. How does the author select the right word, the right phrase, the right image from among the countless conceivable possibilities? How does each person imagine a new existence; reconfigure the choices that make up another destiny?

Sleep on it. Why not? Our dreams contain the infinite. Uninhibited by wakefulness, words and pictures and emotions circulate and combine freely inside our head. Across the centuries, the Unconscious mind has authored some of the greatest works in literature: Goethe and Coleridge are only two of its pseudonyms.

Dreams defy our finite scrutiny; too often they evaporate in the narrow light of day. We are left upon waking only with sweet hints of rain and distant echoes of a song, a nose here, a smile there, some tremor of sadness or flicker of joy, a suggestive and beguiling void. Like a book, like a life, where does the explanation start? A dream has no beginning, and therefore no middle and no end.

I dreamed I entered a house and found its inhabitants all lying on the floor. Lying, but talking and laughing and eating together. Lying instead of sitting. It was like a scene from a book that I had not read and that had not been written. How many such scenes are there to occupy our dreams, our lives, the pages of a book? Infinitely many.

Like Elíasson's sleepwriter, Anton Chekhov faithfully nurtured a little notebook throughout his remarkable career, though we can suppose that he used his mostly during waking hours. Filled with his day-to-day observations of existence's minutiae, the pages preserve glimpses of "ordinary" life's infinite permutations.

"Instead of sheets — dirty tablecloths."

"In the bill preserved by the hotelkeeper was, among other things: 'Bugs — fifteen *kopecks*.'"

"If you wish women to love you, be original; I know a man who used to wear felt boots summer and winter, and women fell in love with him."

This endless variety inspired many of Chekhov's tales. In "The Lottery Ticket," a middle-class couple envisage the potential lives that would follow a jackpot.

> The possibility of winning bewildered them.... "And if we have won," he said, "why it will be a new life, it will be a transformation!"... Pictures came crowding on his imagination, each more gracious and poetical than the last. And in all these pictures he saw himself well fed, serene, healthy, felt warm, even hot!... "Yes, it would be nice to buy an estate," said his wife, also dreaming.... Ivan Dmitritch stopped and looked at his wife. "I should go abroad you know, Masha," he said. And he began thinking how nice it would be in late autumn to go abroad somewhere to the south of France, to Italy, to India!

Writing half a century after his fellow countryman, another precocious note taker, Vladimir Nabokov, composed his novels in two alphabets and three languages (Russian, French, and English). From a blank vivid room, anagrams, puns, and neologisms spilled out. He compared composing a story with fitting together the pieces of a jigsaw puzzle.

> Reality is a very subjective affair.... You can get nearer and nearer, so to speak, to reality; but you never get near enough because reality is an infinite succession of steps, levels of perception, false bottoms, and hence unquenchable, unattainable. You can know more and more about one thing but you can never know everything about one thing: it's hopeless.

Like a jigsaw puzzle, like a dream, Nabokov's novels emerged nonlinearly; he often wrote the middle of a story last. Chapter 8 of a draft manuscript might appear long before chapter 7 or chapter 3. He would frequently write a new story backward, starting out from its final lines.

Lolita, Nabokov's most famous (and infamous) novel, began life on a long series of three-by-five-inch index cards. He sketched out the story's closing scenes first. On subsequent cards Nabokov jotted down not only paragraphs of text but also plot ideas and other bits of information; on one, a chart of statistics on the average height and weight of young girls; on another, a list of jukebox songs; on a third, an illustration of a revolver.

Every so often Nabokov would rearrange his index cards, searching for the most promising combination of scenes. The number of possible permutations would have been immense. Three of Nabokov's cards can be rearranged in a total of six different ways—(1, 2, 3), (1, 3, 2), (2, 1, 3), (2, 3, 1), (3, 1, 2), (3, 2, 1)—while ten cards (equivalent to between two and three printed pages in a book) would be capable of permuting into more than three and a half million sequences. To compose only four or five pages (equivalent to the contents of about fifteen index cards) would require a choice from among some 1.3 trillion variations. *Lolita* runs to sixty-nine chapters and over 350 pages, which means that the number of its potential versions exceeds (by an almost unimaginable margin) the number of atoms that make up our universe.

Of course, many of these potential *Lolita*s would simply not have been viable. And yet among the bewildering, nonsensical, or ham-fisted editions, readable alternatives must exist. How many? A hundred? A thousand? A million? More. Many more. Publishers could produce enough of them to give every reader on the planet his or her very own *Lolita*. In one, the famous opening couplet, "Lolita, light of my life, fire of my loins. My sin, my soul," would appear halfway down page 39—perhaps replaced with a line that Nabokov placed in his chapter 2: "My very photogenic mother died in a freak accident (picnic, lightning)..." In another reader's edition, the couplet shows up at the top of page 117. This *Lolita* begins instead, "I saw her face in the sky, strangely distinct, as if it emitted a faint radiance of its own." In a third version, the original couplet serves as the story's closing lines.

For all I know, some of these incalculably many editions were actually published, each with their subtle yet striking alterations. Perhaps this would explain why the *Atlantic Monthly*'s reviewer called the book "one of the funniest serious novels I have ever read," the *Los Angeles Times* declared it "a small masterpiece...an almost perfect comic novel," and the *New York Times Book Review* announced, "technically it is brilliant... humor in a major key," whereas Kingsley Amis read a book leading to "dullness, fatuity, and unreality" and Orville Prescott, writing for the *New York Times,* found the story "dull, dull, dull."

Which *Lolita* did they read?

It is the writer and reader together who compose their infinite tale. The Argentine writer Julio Cortázar created a novel in which he made this principle explicit. *Rayuela* (*Hopscotch*) was published fifty years ago, not long after *Lolita.* It contains 155 chapters (over some 550 pages), which can be read in two distinct ways. Either the reader starts at chapter 1 and continues reading linearly until the end of chapter 56 (the chapters and two hundred pages that remain being considered "expendable"), or else he begins at chapter 73, then turns back to chapter 1, continues to chapter 2 before jumping forward to chapter 116, then back to chapter 3, forward to chapter 84, and so on back and forth between the chapters according to a "Table of Instructions" at the front of the book.

In one of his "expendable" chapters, Cortázar describes the book's goal:

> It would seem that the usual novel misses its mark because it limits the reader to its own ambit; the better defined it is, the better the novelist is thought to be. An unavoidable detention in the varying degrees of the dramatic, the psychological, the tragic, the satirical, or the political. To attempt on the other hand a text that

> would not clutch the reader but which would oblige him to become an accomplice as it whispers to him underneath the conventional exposition other more esoteric directions.

As Cortázar's accomplice, we follow the novel's hero—an Argentine bohemian—through the streets of Paris as he contemplates his life and its inexhaustible potential paths. We begin the book either at chapter 1, like this: "Would I find La Maga?" Or else at chapter 73, on page 383: "Yes, but who will cure us of the dull fire, the colorless fire that at nightfall runs along the Rue de la Huchette..."

Turning the pages, reading different stories. For instance, the reader who begins at chapter 1 will soon reach this line in the fourth chapter: "[She] picked up a leaf from the edge of the sidewalk and spoke to it for a while." For the other reader, however, "chapter 4" is really the seventh chapter of the story, preceded by the sixth chapter, which is labeled "chapter 84." In this chapter, on page 405, he reads, "I keep on thinking of all the leaves I will not see, the gatherer of dry leaves, about so many things that there must be in the air and which these eyes will not see... there must be leaves that I will never see." These lines enrich the second reader's understanding of the woman who, several pages later, on page 25, will pick up a leaf from the edge of the sidewalk and speak to it.

A consequence of reading in this way is disorientation; the leapfrogging reader lacks any sense of having completed the book. He reads the final lines on the final page of the physical book long before he comes to any conclusion of the story. Arriving later at the one hundred and fifty-third chapter (labeled "chapter 131"), he proceeds to the following chapter (labeled "chapter 58"), only to discover that he should return again to chapter 131. An interminable loop between the two "final" chapters appears. What is more, assuming he has kept count, the reader notices that the chapters — read in this order — total 154. One of the chapters — "chapter 55" — is absent from the list.

Hopscotch's structure demands that readers make their own sense of the story. One might decide to read the chapters consecutively, but in descending order, starting at chapter 155. Another decides to read all the even chapters before the odd: 2, 4, 6, 8 . . . 1, 3, 5, 7 . . . A third does the reverse, reading all the odd chapters before the even. A fourth reads only the prime-numbered chapters: 2, 3, 5, 7, 11, 13, 17, 19, 23, 29, 31 . . . finishing at chapter 151 (thirty-six chapters in all). A fifth begins at the first chapter, then reads the third (1 + 2), turning next to the sixth (1 + 2 + 3), followed by the tenth (1 + 2 + 3 + 4) and so on.

Just when the plucky reader has attained the end of one story, another story beckons him to pick up its pages and start again. The book of ascending chapters becomes a book of descending chapters. The book of odd-number chapters becomes a book of even-number chapters. Every reading differs; every reading offers something new. It is impossible to dip into the same book twice.

I am reminded of Nabokov's view that we can never read a book: we can only reread it. "A good reader, a major reader, an active and creative reader," says Nabokov, "is a rereader." Initial readings, he explains, are always laborious, a "process of learning in terms of space and time what the book is about, this stands between us and artistic appreciation."

Think of the countless stories of Chekhov, of the innumerable editions of *Lolita* and *Hopscotch,* which lie before every reader's eyes unnoticed, unloved, unread.

Flaubert, in a letter to his mistress, wrote, "How wise one would be if one knew well only five or six books." It seems to me that even this figure is an exaggeration. To learn infinitely many things, we would only ever need perfect knowledge of one book.

Eighteen

POETRY OF THE PRIMES

Arnaut Daniel, whom Dante praised as *"il miglior fabbro"* ("the better craftsman"), sang his love poems in the streets of twelfth-century southern France. Of his life not much is known, but I find it tempting to link a brief and rare report about the troubadour with the *sestina* form of poetry (six stanzas, each containing six lines, plus a concluding half stanza) that he invented.

Raimon de Durfort, a contemporary of Arnaut, called him "a scholar undone by dice." This alleged acquaintance with gambling suggests a possible influence for the sestina's form. A die, as everyone knows, has six faces. The throw of a pair of dice creates a range of outcomes amounting to thirty-six, which is the total number of lines in the poem's six stanzas. So far as I can tell, no one has made this connection between the sestina and the die before—maybe because it is a connection little worth making. I leave it for the reader to decide.

Unusually, the sestina does not run on rhyme, symbolism, alliteration, or any other of the poet's typical devices. Its power is the power of repetition. The same six words, one at the tail of each line, persist and permute across every stanza (in the final half stanza, the words appear two to a line) according to an intricate pattern.

First stanza: 1 2 3 4 5 6

Second stanza: 6 1 5 2 4 3

Third stanza: 3 6 4 1 2 5

Fourth stanza: 5 3 2 6 1 4

Fifth stanza: 4 5 1 3 6 2

Sixth stanza: 2 4 6 5 3 1

Conclusion: 2 1, 4 6, 5 3

Which is to say, the final word in line six of the first stanza (1 2 3 4 5 **6**) reappears as the last word of the next stanza's opening line (**6** 1 5 2 4 3), and at the close of the second line of stanza three (3 **6** 4 1 2 5), and so on. Here follows an example, from Dante (employing the six words "shadow," "hills," "grass," "green," "stone," and "woman").

I have come, alas, to the great circle of shadow,
to the short day and to the whitening hills,
when the color is all lost from the grass,
though my desire will not lose its green,
so rooted is it in this hardest stone,
that speaks and feels as though it were a **woman.**

And likewise this heaven-born **woman**
stays frozen, like the snow in shadow,
and is unmoved, or moved like a stone,
by the sweet season that warms all the hills,
and makes them alter from pure white to green,
so as to clothe them with the flowers and grass.

When her head wears a crown of grass
she draws the mind from any other **woman,**
because she blends her gold hair with the green
so well that Amor lingers in their shadow,
he who fastens me in these low hills,
more certainly than lime fastens stone.

Her beauty has more virtue than rare stone.
The wound she gives cannot be healed with
grass,
since I have traveled, through the plains and hills,
to find my release from such a **woman,**
yet from her light had never a shadow
thrown on me, by hill, wall, or leaves' green.

I have seen her walk all dressed in green,
so formed she would have sparked love in a stone,
that love I bear for her very shadow,
so that I wished her, in those fields of grass,
as much in love as ever yet was **woman,**
closed around by all the highest hills.

The rivers will flow upwards to the hills
before this wood, that is so soft and green,
takes fire, as might ever lovely **woman,**
for me, who would choose to sleep on stone,
all my life, and go eating grass,
only to gaze at where her clothes cast shadow.

Whenever the hills cast blackest shadow,
with her sweet green, the lovely **woman**
hides it, as a man hides stone in grass.

An air of expectancy permeates the text: since the reader knows what is coming, the poem must rise to the challenge of delivering surprise. The sestina plays with meaning, conferring new aspects, in changing contexts, on the same word. A tension between the law of the numerical pattern and the liberty of the author is ever present, ever palpable.

Artists and mathematicians alike have been drawn to the sestina's numerous properties. In their wonderful book, *Discovering Patterns in Mathematics and Poetry,* the mathematician Marcia Birkin and poet Anne C. Coon compare the rotation of words in a sestina to the shifting digits in a cyclic number.

Cyclic numbers are related to primes. Division using certain prime numbers (such as 7, 17, 19, and 23) produces decimal sequences (the cyclic numbers) that repeat forever. For example, dividing one by seven (1/7) gives the decimal expansion 0.142857142857142857... where the six digits 142857—the smallest cyclic number—continue around and around in a never-ending ring dance.

If we multiply 142,857 by each of the numbers below 7, we see that the answers are permutations of the same six digits.

$1 \times 142857 = 142857$

$2 \times 142857 = 285714$

$3 \times 142857 = 428571$

$4 \times 142857 = 571428$

$5 \times 142857 = 714285$

$6 \times 142857 = 857142$

In this instance, the digit 7 at the end of the first answer (142857) reappears in the fourth position of the second answer (285712) and in the fifth position in the third (428571) and so on. Each digit rotates through every answer, changing place at every turn, like the end words in the stanzas of a sestina.

Hazard has no place in the sestina. The end words in every stanza fall at once into line, the position of each determined before the poem begins. Algebraically, we can describe the sestina's structure (from the second stanza onward) like this.

$\{n, 1, n-1, 2, n-2, 3\}$ where n refers to the number of stanzas (six).

How our medieval troubadour concocted this clever pattern is unknown. His deep familiarity with the rhythms of words and music likely helped. In one of his few surviving songs he says:

> Sweet tweets and cries
> and songs and melodies and trills
> I hear, from the birds that pray in their own language,
> each to its mate, just as we do
> with the friends we are in love with:
> and then I, who love the worthiest,
> must, above all others, write a song contrived so
> as to have no false sound or wrong rhyme.

Of course, that Arnaut plumped for six stanzas, and not five or seven, probably owes as much to chance as the outcome of the toss of a die. In fact, a small number of poets have tried their hands at *tritina*s (containing three stanzas) and *quintina*s (containing five), with some success. Raymond Queneau, a French poet with a mathematician's itch for generalization, eager to understand how the pattern works, explored the limits of the form. In the 1960s, he worked out that only certain numbers of stanzas permitted the permutations of the sestina. A four-stanza poem, for example, produced jarring alignments of the same word.

{n, 1, n –1, 2}

First stanza: 1 2 **3** 4

Second stanza: 4 1 **3** 2

Third stanza: 2 4 **3** 1

Fourth stanza: 1 2 **3** 4

The same went for a poem of seven stanzas.

{n, 1, n –1, 2, n – 2, 3, n – 3}

First stanza: 1 2 3 4 **5** 6 7

Second stanza: 7 1 6 2 **5** 3 4

Third stanza: 4 7 3 1 **5** 6 2

Fourth stanza: 2 4 6 7 **5** 3 1

Etc. . . .

After much trial and error, Queneau determined that only thirty-one of the numbers smaller than 100 produced the sestina's pattern. His observation led mathematicians to discover a surprising relationship between the sestina and the primes. Namely, poems containing three or five stanzas behave like the six stanzas in a sestina, because 3 (or 5, or 6) × 2, + 1, always equals a prime number. For the same reason, sestina-like poems of eleven, thirty-six, or ninety-eight stanzas are all possible, but not those containing ten, forty-five, or one hundred.

Sestinas are not the only form of poetry to be shaped by primes. Brief and glancing, haiku poems also derive their strength from these numbers.

The Japanese have long been disposed to brevity. Inquiries for the name of "Japan's Shakespeare" or "Japan's Stendhal" will be, in the best of cases, greeted with blank stares. Oriental epics fell into complete neglect at about the same time that the Viking Snorri Sturluson was putting the finishing touches to his saga. Courtiers of the Heian period (dating from the eighth to the twelfth centuries, a period that the Japanese consider a high point in their history) worked up the most extended pieces by concatenating dozens of short verses, by many hands. Dignitaries alone, however, had the right to commence these chain poems by inventing their opening three lines (called the "hokku"). Among images of romantic love and soul searching, these opening lines always contained a reference to the seasons, and an exclamation such as *ya* ("!") or *kana* ("how . . . ! what . . . !"). However, even this carefully wrought convoy of miniature verses became too cumbersome in the end for Japanese tastes, so that generations of mouths gradually eroded them to the triplet of lines that we know today as haiku.

Like the sestina, the haiku uses no rhyme. Its three lines contain five, seven, and five *onji* (syllables), seventeen in all. Three, five, and seven are the first odd prime numbers. Seventeen, too, is prime.

One possible (if partial) explanation for this structure is the marked Japanese preference for odd numbers. In the annual *Shichigosan* (Seven-Five-Three) festival, three-year-old children of both sexes, five-year-old boys, and seven-year-old girls visit shrines to celebrate their growth. Cheerleaders at a sports match clap in three-three-seven beats. Even numbers, meanwhile, are virtual bogeymen. The number two represents parting and separation, while four is associated with death. In several phrases the number six translates roughly as "good-for-nothing."

Prime numbers contribute to the haiku form's elemental simplicity. Each word and image calls out for our undivided attention. The result is an impression of sudden, striking insight, as if the poem's objects had been put into words for the very first time. A sample from the form's most celebrated practitioner, the seventeenth-century poet Matsuo Bashō, illustrates this.

Michinobe no (The mallow flower)
Mukuge wa uma ni (Against the side of the road)
Kuwarekeri (Eaten by my horse)

There exists as well a slightly longer version of the haiku, the *tanka* form, which complements the haiku's trio of lines with two further lines of seven syllables each (known as the *shimo-no-ku,* or "lower phrase"). The *tanka*'s total number of syllables—thirty-one—is, once again, a prime.

Bashō, whose name today is synonymous with the haiku form, counted various influences on his work, but chief among them was a wandering monk who lived at the time of Arnaut Daniel and wrote some of the finest *tanka* verse. His name was Saigyō. Something of the master's evocative simplicity is suggested in the following *tanka.*

Michi no be ni (On that roadside lea)
Shimizu nagaruru (Where pure, crystal waters flowed)
Yanagi kage (Grew a willow tree)
Shibashi to te koso (For a little while I stayed)
Tachidomaritsure (There and rested in its shade)

It seems that the image of Saigyō's willow tree took root in the imagination of many generations of poets. Some five centuries after the poem's composition, Bashō underwent a pilgrimage to its site in the north. In his travel diary he noted that "[t]he willow about which Saigyō wrote the famous poem still stood by a rice field in Ashino village. As Mr. Koho, who governed this county, always wanted to show the tree to me, I had been anxious to discover its location. I was happy to stop by the tree today."

Bashō's homage culminated in a haiku dedicated to Saigyō's tree.

Ta ichimai (Over a whole field)
Uete tachisaru (They have planted rice, before)
Yanagi kana (I leave the willow)

As I think of the complicity between poems and primes, perhaps the only surprise is that we should even find it surprising. Viewed one way, the relationship makes a perfect kind of sense. Poetry and prime numbers have this in common: both are as unpredictable, difficult to define, and multiple-meaning as a life.

This lifelike quality connecting them is too often overlooked. Many poems, it is true, lie mothballed in slim anthologies; many primes languish in mathematicians' sums. Picked and pored over by their experts, they lose the public's attention (and affection) for the academic company they keep.

And yet we see her so clearly, Dante's woman, as she wanders through our memory, and Bashō's horse, munching his flower, looks and sounds all too real. Free (as a prime number) from a rhyme's reassurance or a storybook's rules, the images swerve and forestall our expectations—and keep all clichés at bay.

Poems and primes are tricky things to recognize. A glance will usually not suffice to tell us if such-and-such a number has factors, or whether a given text contains much meaning. Even old hands can find it hard to tell genuinely felt verse from a trite list of nice-sounding words, or spot a composite number as being merely a pastiche of lesser primes.

Dante's sestina, haiku verses, and the hither and thither of the primes—for each we ask ourselves, what does it mean? Are we, in the end, any closer to the woman whose beauty "has more virtue than rare stone"? Her face changes with the stanzas, offering us a multiplicity of perspectives. And what of Saigyō's willow tree, which provides at once shade and reflection?

The same is true of the prime numbers—an ancient mathematical mystery. Thirty-one, the number of syllables in the *tanka,* is a twin prime (being two apart from its neighbor twenty-nine), and a Mersenne prime—being one less than a power of two: $(2 \times 2 \times 2 \times 2 \times 2) - 1$—but such labels fall far short of explanation. That is because we do not really know why the prime numbers appear where they do. Many unproven conjectures remain. The reader of poetry and the mathematician are left finally only with hints and fragments, minus any big picture—as in life.

Nineteen

ALL THINGS ARE CREATED UNEQUAL

Unlike diabetes or curly hair, poverty rarely skips a generation. A parent's bank balance will often dictate his child's destiny to a far greater extent than his blood. Blond mothers sometimes produce brunet babies; tall men may not always spawn basketball pros; but more than ninety times out of every hundred, the poor beget more poor.

I am the son of poor parents, poor grandparents, poor great-grandparents, and so on. Suffice it to say that more than a few grisly tales of hard times have come down to me. One of these was told to me by my father, back in the early 2000s, not long after I had flown the family nest to serve my apprenticeship in the adult world. I was living in a house south of London that belonged to someone else. It was the first time I had shared with a roommate. The place had little to commend it. Small, out of the way, it was very modestly decorated. In my box of a bedroom, I slept on a sofa bed, pale green, the color of a plant in shade. I subsisted on students' fare: small plates of pasta, sandwiches, and beans on toast.

Some evenings I took calls from home. On one occasion my father rang, and we fell into a long conversation. His chattiness, I must admit, took me rather by surprise. He had never been one for self-disclosure. Why then was he opening up to me? For an eldest son's compassion, for a walk accompanied down memory lane? I do not know. We had been talking about nothing in particular when all of a sudden he said, "We moved around a lot when I was a kid."

"Sorry?"

"My parents, well, my mum and her man. It is a long story."

And, just like that, in a matter-of-fact voice, he told me his tale. He told it with such simplicity and such precision, not a word or image out of place, that it dawned on me only later that he had certainly long rehearsed this moment in his mind. Even as I listened I knew better than to interrupt him with my questions or comments. He spoke, as we say, "from the heart"; this man, my father, telling me, his son, such things as he considered important and valuable for me to hear.

One scene in particular struck me. As a boy of ten or thereabouts, he and his parents had been returning home one summer evening from a trip to the town fair. The front lawn, they noticed as they approached the house, had been eerily altered. Running ahead, my father was stunned by what he found. Chairs and tables and pots and pans and beds and lamps had been evacuated. The furniture huddled destitute in a big pile. A padlock on the front door barred their entry.

I might have asked, "Why had the landlord not given my grandparents more time to clear their debt?" But I did not ask. The thought of the furniture left out in the front yard made a deep impression on my mind. It was the image of a home turned inside out, its intimacy smashed, its innards spewed. How terrible they must have looked, those useless lamps and long bare table legs and the embarrassing tea-colored letters in a half-open desk drawer. The scene felt so vivid that it made my eyes smart.

My father was born in 1954, the year that followed Queen Elizabeth's coronation. My grandparents' expulsion from their home happened ten years later—in the midst of the swinging sixties, the decade of "peace and love." In their book *The Poor and the Poorest,* published in 1965, the sociologists Peter Townsend and Brian Abel-Smith estimated that this ten-year period, the first ten years of my father's life, had produced a near doubling in the percentage of Britons living beneath the breadline, rising from 8 percent to 14.

In 1979, the year of my birth, Peter Townsend published a further study, *Poverty in the UK,* showing that relative hardship stifled the lives of 21 percent of the population. This figure appears to have remained more or less stable ever since.

A final statistic, this one from 2008, describing the generation after mine: according to the London School of Economics, the household wealth of the top 10 percent of the population now towers one hundred times above that of the lowest 10 percent.

Inequality is invidious. It is also universal. No country, to judge from the comparative data, has been spared it. Every land has its share both of hovels and five-star hotels. Talk of a "classless society" has, time and time again, proved to be nothing but hot air. Western sympathizers who visited the Soviet Union in the 1930s were disappointed to find that its October Revolution had hardly "abolished" the gulf that divided rich and poor. Zero in imperial units, they learned, was still zero under the metric system. Meanwhile beneath their mud-brown uniforms, stiff with starch, the Kremlin's rulers continued to wear the slaughtered emperor's clothes.

But enough with statistics: I wonder if mathematics can do more about the phenomenon of disparity than simply measure it. I wonder if it can tell us something about what kind of thing disparity is: where does it come from? What makes it grow or shrink? Can mathematical thinking address these kinds of questions?

It can. Mathematics and money both originated in abstraction. As with mathematics, we owe the concept of money to the ancient Greeks. They who first abstracted "five" from the fingers on one hand, first stamped "five drachmas" on a metal coin. And just as the concept of "five" slipped from the fingers that had described it, becoming applicable to anything—men, crumbs, daydreams—with identical quantity, so a coin's "value" exceeded the metal of its composition, capable of transforming into anything agreed as having equal worth.

Substituting numbers for objects changed the world, for better or worse. At once everything became quantifiable, even the light of the moon, which Aristophanes in one of his plays describes as saving people a drachma's worth of torches per month. Where previously, bartering and the exchange of gifts had settled all Athenian transactions, now most social dealings came down to sums. Reciprocity between citizens gave way to the potentially unlimited accumulation of individual "wealth." We read with a feeling of familiarity Aristotle's lament (in his *Politics*) that some doctors turn their skills into the art of making money. Sophocles goes much further, placing in one of his character's mouths a blistering denunciation of money as that which "lays waste even cities, expels men from their homes," and "thoroughly teaches and transforms good minds... to know every act of impiety."

In its abstraction, money acquired the impersonal neutrality of numbers. Goods no longer embodied the donor's generosity or personality; calculation replaced feelings. Individual autonomy grew, but so too did an egoism that made money the measure of all things. And like abstract numbers, money became invisible. Coins might be concealed far more easily than cows. Lycurgus, ruler of Sparta, found that he could only fight the "injustice" of the rich hiding away large sums only by making his iron coins so large and so heavy that ten of them alone required transportation upon a wagon.

Because numbers can go on forever, money has no limit. Of wealth, Aristophanes tells us, a person can never have enough. Bread and sex and music and courage all sate the preceding appetite, but wealth does not. It is impossible to put a check on moneymaking. If a man receives thirteen coins, he will hanker after sixteen, and possessing them he considers life unbearable unless he now earns forty. Nature, it might be observed, imposes strict boundaries on a person's height and age span, so that in even the most extreme instances no one can rise or fall too far or too short from the rest, whereas no such boundary inhibits money. Think of King Croesus who had so much gold that he gave it away. It was Croesus whom Solon the legislator memorably warned when he said that he who has much, has much to lose.

I am reminded here of the story I first heard on the morning radio two years ago, of an elderly Parisian heiress who had been flattered into great feats of generosity by a much younger man. Of course, the story immediately spread to print. I read the fine details across consecutive pages of my daily paper. The billionaire had allegedly been pampered out of paintings by Munch, Picasso, Matisse, seduced into gifting precious tomes and manuscripts and, all in all, fussed over to the tune of scores of millions of euros in handouts over several years.

Poor thing!

Solon, I should mention, became the first man in history to make laws addressing inequality. We read in Plutarch how the mass of ancient Athenians in Solon's day found themselves in hock to the city's wealthy aristocrats. Some had been sold into slavery or else had handed over their children as collateral, while others had fled with their family into exile. When he was elected chief magistrate, Solon promptly divided the city's populace into categories, granting to each category proportional responsibilities and rights. The highest class consisted of those earning an income in excess of five hundred bushels. In the second class were those citizens able to afford a horse (and therefore to pay a "horse tax" upon the purchase). Yoke-of-oxen men comprised the third class, with an annual income between two hundred and three hundred bushels. Freed from the fear of enslavement, the remainder of landless citizens could attend public assemblies for the first time, and sit on juries.

We can express the distribution of wealth in any society as a formula, using a number *x* between 50 and 100 such that *x* percent of the society's wealth belongs to (100–*x*) percent of the population. In a highly egalitarian society (where *x* equals, say, 55 or 60), 45 percent (or 40 percent) of the people would hold 55 percent (or 60 percent) of the assets. Most Western societies, however, show a far more skewed distribution. Economists have found that *x* in most developed countries equates to a number in the region of 80, meaning that 80 percent of the society's fortune pads the pockets of only 20 percent of its members.

Naturally, the share will vary from person to person. Money is fickle, always changing owners. That different people have dissimilar wealth is not surprising. What surprises is the scale and constancy of the divide. The economist and mathematician Vilfredo Pareto, who first observed (at the end of the nineteenth century) that 20 percent of Italians owned 80 percent of the nation's wealth, found nearly identical results when he studied the historical data from many other parts of Europe. The distribution of wealth in Paris since 1292, he discovered, had hardly moved at all. Later researchers confirmed these findings.

Because most men and women, for want of resources, remain at the bottom, the elite, for want of competition, remain at the top. The poorest expend all their energies simply to keep body and soul together. I think of Degas's painting of the two peasant women ironing: one is anonymous, hunched over her iron; the other yawns candidly, her mouth the shape of an "O." The yawn distorts her features, undoing the individuality of her face.

The Spartan Lycurgus, we saw, made money as big as men; imagine, an instant, men as big as their money. Picture the difference between a miller and a millionaire. The man who mills grain owns perhaps no more than one-thousandth of the rich man's fortune: the millionaire, his advantage converted into height, would be a thousand times as tall. To him, the miller would appear no bigger than an ant. With whom will the giant do business? Only with someone big and strong enough to carry his employer's burden. Likewise this someone, smaller than the millionaire on whom he is dependent, but still far greater in size than the grain-running ant, to whom will he entrust his dealings? To his peers. They, in turn, will do the same. Good manners and compromise characterize the bulk of these transactions, but hardly anyone thinks to condescend to those with whom they have next to nothing in common. Our lowly ant friend is simply nowhere in sight.

Every man and woman, at whichever point along the scale, prefers to look up, not down. Even the miller gives a hand to his equal or his superior, rather than to someone far below him, for fear of conceding his rank to the worse-off man. With the wealthier, he is generous; with the poorer, he is mean. He has little, but the little he has goes purely to keeping up his minor station.

Comparisons are somewhat facile, of course. Beside a billionaire, even the millionaire is poor. The world's hundredth-richest person has but one dollar for every eight in the pocket of the world's richest man.

Whether the economy expands or whether it shrinks, the obsession with keeping up one's station remains. The inequality this obsession feeds off is a precocious learner: the more of it there is, the faster it will grow. Take the hypothetical egalitarian society, for example, where 45 percent of the population own 55 percent of its wealth. In such a society, around 20 percent (45 percent of the 45 percent) of citizens own about 30 percent (55 percent of the 55 percent) of the total resources. By the same logic, a sixth (55 percent of the 30 percent) of the society's goods belongs to one citizen in every eleven (45 percent of 20 percent).

The contrast between this theoretical society and most of our modern cities—those that obey Pareto's 80–20 principle—is striking. Wealth in these places can propagate far more ruthlessly, dramatically: the accounts of as few as four individuals in every hundred (20 percent of 20 percent) will bloat with as much as two-thirds (80 percent of 80 percent) of all the available income. And of these four men made of money, the very richest might have up to one-half (80 percent of the two-thirds) all to himself.

Human beings and their self-interest are inseparable, but inequality needs a society to invent it. The creation of any vast and ambitious social project demands an unequal allocation of resources in order to achieve its goals. Without substantial inequality, as John Maynard Keynes pointed out, Europe's railroads—a "monument to posterity"—would never have been constructed. Tolstoy, for one, hated the railroads precisely because they represented this inequality, going even so far as to throw his best-loved character under a murderous train. It is true that most of the men who laid the rails never had the opportunity to ride them. Why then did they agree to do so? The railroad workers chose to cooperate with the wealthy, Keynes argued, on the tacit understanding that what they produced together, using the money of one side, the labor of the other, would ultimately serve the nation as a whole, and the principle of "progress." World war, however, would subsequently smash this fragile alliance between the classes, by shaking the faith of both in the future. The bludgeoning of bombs and the gutting of gunfire disclosed to all "the possibility of consumption... and the vanity of abstinence."

When Keynes spoke of the value of inequality he did not mean unbridled inequality. He meant an inequality that was consensual and that served a collective purpose. The selfish motive of making money, he admitted, helped to produce goods and services that benefit many. The same motive could also turn certain "dangerous human proclivities"—cruelty, self-aggrandizement, and the tyrannical pursuit of power—to more harmless pursuits. But, he was not at all complacent.

> It is not necessary for the stimulation of these activities and the satisfaction of these proclivities that the game

> should be played for such high stakes as at present. Much lower stakes will serve the purpose equally well, as soon as the players are accustomed to them. The task of transmuting human nature must not be confused with the task of managing it.

How we lower those stakes is a question I leave to our politicians. I will not hold my breath. There is no easy solution—none analogous to a politician's promise. Money's abstractness is complex, evasive. It turns our world upside down. For instance, a farmer's cow calves more prodigiously than other cattle—this is normal, part of nature. What, though, should we think of houses that beget more houses? Like many of those born poor, my father never had the chance to own the roof above his head, whereas the proprietor of four houses will likely wind up with six, or twelve, or twenty.

All of which brings us to the question that Tolstoy posed his readers in the short story "How Much Land Does a Man Need?" Pahom, the greedy peasant, kills himself in the endless pursuit of more and more acres.

"They are poor in the midst of riches," Seneca observed of some of ancient Greece's wealthiest—and greediest—men, "which is the worst kind of poverty."

Twenty

A MODEL MOTHER

Not long ago, my mother's age reached double mine. Two lives when compared with my young man's span—half of her that I cannot see.

My mother has always been a mystery to me. We have had my lifetime to get to know each other, but it still feels nowhere near long enough. Her behavior eludes me; it outpaces my powers of comprehension. Try as I might, I cannot figure her out.

Her face has not changed much with the years. She often wears an expression that flits between smugness and fright. The same stubborn creases, made by steady clenching, around the mouth; the same defiant twinkle in her eyes. She smiles fitfully, unexpectedly, as if bestowing a favor. Beneath the fine graying hair and wrinkles, I can still find myself in her gaze.

Memories. Take the kitchen of my childhood, for example, where my mother would spend the lion's share of her day. I remember her prowling the linoleum, pen and notepad in hand, compiling a list of groceries to buy. She was alive to every sound and suspicious of the slightest incursion. She poked her nose into cupboards and the fridge, looking for the cans of beans and the bottles of milk and the packets of cheese and sliced bread bought only a day or two before. Thin air had taken their place.

"The children eat us out of house and home," she complained to my father.

My father assumed a posture of resignation.

We kids knew which of our parents held command. Or at least, we thought we did. At other times my mother would suddenly come over all shy and prone to blushing. My father could hardly draw a word out of her.

Then there were the Christmas presents. Year-round she would ferret out bargains at local yard sales, sequestering the toys and games in closets or under beds until Santa's sleigh arrived. Of course, we always knew where to find them, but in the spirit of the season we turned a blind eye. It was never easy: buried treasure seemed to lie in every nook and cranny of every room. Why, then, many Decembers later, under piles of old clothes, did we bring unopened presents to the surface? Had she simply mislaid them, forgotten their location? Could it really be that the buying of them had meant more to her than the offering?

A mathematician would say, "Graph the data." That is how mathematicians speak. And it is true enough: puzzling occurrences usually require the long view—and a firm grasp of context. In my preteens I decided that if only I could assemble enough of my observations, and settle on some parameters for their analysis, it might be possible to make a predictive model of my mother's behavior.

It was around the time that I began to look like her, once the elementary school blackboard became a blur. Our myopia, it could be said, brought us closer together. "A mommy's boy," my father sometimes called me. None of my brothers earned this epithet. Spending more and more time in her company, I felt the puzzle of her presence intensely.

Back then, a thigh thinner, she was always on the go. I took to mapping her movements. On Saturday mornings my mother would return from the local library, carrying a couple of romance paperbacks in her arms, smelling faintly of must. In the space in the living room before the television, I would sit for what seemed like hours half listening to the yellow rustling of the pages that originated from the settee. Every other Sunday, robed in her best dress, she would take us—my brother, sister, and I—to the neighbor's on the corner to drink tea and trade gossip. Midweek, she would venture out to do the rounds of secondhand shops, returning with bags fattened on goods that nobody seemed to need.

Perhaps she noticed me tracking her and wished to catch me out, or perhaps she simply grew bored with her own repetition, but for whatever reason she would sometimes decide to mix it all up. Come Saturday, the living room would contain the same atmosphere. But now, dog-eared biographies exhaled the musty smell. Without warning, the front door would stay shut for the Lord's Day of rest; we took tea with the neighbor an evening after school instead. Even her favorite shops suddenly became good only for returning things.

One afternoon she took me with her to return a pair of shoes. Approaching the shop, I compared the mother in my head with the real thing. The imaginary mother would choose a male shop assistant (my mother, I knew, hated to haggle with other women). She would complain that the shoes pinched her young son's toes, and when the man lifted the offending footwear from its container, she would add that the leather scuffed immediately. To the request for a receipt (which she always lost), her voice would rise in volume, detailing all the children's feet that she had to keep dry and warm. The man would then nod patiently before offering an exchange.

Unfortunately, on this occasion my real mother bore no resemblance to her model.

A young woman with tight bound hair assisted all the customers. My mother's voice as she handed over the shoebox sounded weak, her words unacquainted with one another. "I'm sorry to hear it," the woman interrupted, as though offering her condolences. "I'm sorry to hear it, but there really is nothing that I can do."

I confidently expected my mother to put up a fight. But she promptly slumped into one of the seats intended for the trying-on of shoes, with only a long sigh for a rejoinder. The shop assistant reiterated that a refund was quite out of the question. My mother looked down at the floor and merely sighed again. Eventually, when my mother showed no signs of relenting, the young woman said, "Please leave," and then, "Please leave, or I will have to call the police." My mother sagged even deeper into her seat and crossed her legs.

My ten-year-old self was filled with apprehension. My imaginary mother would never have behaved like that! It took a long time for me to understand my real mother's sit-in. Of course, she knew all too well what she was doing. She knew that the young lady had conceived her own version of my imaginary mother. On that day, in that shop, my mother in her flesh and blood defied them both.

At last, driven to exasperation, the woman pulled something small and shiny from her pocket. "Do not breathe a word of this to anyone," she said, before sketching a long gash along one of the shoes with the blade of her penknife. "This way we get a refund from the manufacturer for damaged goods."

In fact, it did not bother me as much as you might think when the actions of my mother and her imaginary doppelgänger failed to coincide. I slowly came to understand how limited and clumsy an approximation was my model of her, how many variables I had not accounted for (whose existence I had not even guessed), and how large and liberating a role chance played in all our affairs. Besides, each variation between the imaginary and the real provided me with new clues. This variation, I hoped, growing larger or smaller by turns, would act as a sort of compass toward a greater understanding of my mother's true nature.

On those rare occasions when my mother became her model's mimic, the eerie sense of déjà vu nauseated me. I worried that it suggested some dark cunning within me, or, worse, deterioration in the liberty of my mother's will. Besides, how could I even be sure that my success showed real merit? Perhaps it could be put down to nothing more than luck: even a stopped watch gives the correct time twice a day.

Possibly, my inability to understand my mother is the result of an anterior uncertainty, that is: what should a mother's behavior look like? I am not talking about an ideal mother—I do not believe, alas, in such creatures—but rather the most common and distinguishing maternal qualities. A baseline, if you will.

This is more difficult than it sounds. For one thing, the category of "mother" admits all manner of members. You can be a mother at sixteen or at sixty (the latter with the help and expertise of scientists); with a single child, or my mother's nine. According to the dictionary definition, a mother is any "female who has given birth to offspring." This is about as roomy and heterogeneous a group of people as we can find. It is, in the words of statisticians, too large a sample.

How, then, should I go about assembling a more workable sample of my mother's peers, one that provides a realistic context for analysis? By looking at mothers of nine? I am not sure that Britain can boast many nine-child families, not since the days when Queen Victoria had her own royal brood of nine. Newspaper articles turn up only a couple of examples: one is a philosophy graduate and boardroom executive who thought she and her Buddhist husband "would stop at five"; the second, a former anorexic who says she feared her chances of getting pregnant were "very, very slim." Naturally, this tiny cohort of women is no more representative of maternity than the first.

I might try a slightly different, related question: what does my mother's behavior tell us about her? But here we run into similar difficulties. For each of her acts, we can imagine a hundred more or less plausible reasons. A hundred imaginary mothers would slug it out for vindication. But each act is the offspring of another: each imaginary mother would be capable of producing one hundred more. Clearly, this approach will take us nowhere closer to any answer. For even if we could somehow locate the "right" reason for each of my mother's actions, and thus identify the "right" imaginary mother within our galaxy of imaginary mothers, we would be left in the end only with an identical twin—as complex, mysterious, and baffling a woman as the one who raised me.

More empirical observation and less abstract reasoning seem necessary, if I am to arrive at any conclusions about who my mother really is. I fully admit that in this I am saying nothing new. When the psychiatrist Édouard Toulouse resolved to objectively measure Émile Zola's genius, for example, he was entirely typical of his times. He took the novelist's height, wrapped a tape around his shoulders, skull, and pelvis, evaluated the strength of his grip, the acuity of his nose, ears, and vision, grilled his powers of recollection and noted the hours when he ate, slept, and wrote. Zola's pulse, the doctor found, was sixty-one before he first put pen to paper, dropping to fifty-three when he called it a day.

Scientists in the Soviet Union also went in for this sort of thing, counting the subject's words rather than his pulse. In their experiments, they tried predicting the next word in a sentence from those that had come before. Young girls in conversation, they discovered, were easiest to anticipate; newspaper columnists followed close behind, while poets proved hardest to second-guess.

Did this result surprise the scientists? They do not tell us. Perhaps the poets took the same liberties with their speech as with their pen. The best poems, the mathematicians determined, combined in equal parts the predictability of meter with the novelty of unusual words. Too much meter made a poem banal; too much freewheeling, on the other hand, rendered it hard to follow. The delicate balance of convention and invention gives meaning to what we say.

The lesson we can learn from these experiments is small but valuable. Mutual understanding depends on our powers of prediction, though they frequently operate beyond our control. With his microscope the psychiatrist got no nearer to what moved Zola's pen, yet he knew intuitively how to talk his old friend into his tests. The Soviet mathematicians could not accurately preview the poets' inspiration, but their conversations once outside the laboratory ranged as far and wide as with anyone else.

In Edgar Allan Poe's story "The Purloined Letter," we see a boy observing and then outguessing each of his schoolmates at a game of marbles. The game consists in determining whether the number of marbles concealed in the opponent's hand is odd or even. For every correct guess, a marble is won; for every wrong guess, one marble is yielded. Thanks to his "astuteness," the boy finishes by winning all the marbles in the school. The boy, Poe explains, makes an intuitive assessment of his rival.

> For example, an arrant simpleton is his opponent, and, holding up his closed hand, asks, "are they even or

> odd?" Our schoolboy replies, "odd," and loses; but upon the second trial he wins, for he then says to himself, "the simpleton had them even upon the first trial, and his amount of cunning is just sufficient to make him have them odd upon the second; I will therefore guess odd";—he guesses odd, and wins.

When the boy considers his opponent "a simpleton a degree above the first," he reasons, "this fellow finds that in the first instance I guessed odd, and, in the second, he will propose to himself upon the first impulse, a simple variation from even to odd, as did the first simpleton; but then a second thought will suggest that this is too simple a variation, and finally he will decide upon putting it even as before. I will therefore guess even." He guesses even, and wins.

Poe goes on to tell us how the marble winner is able to intuit the thoughts and feelings of the boy opposite him: he closely watches and mirrors the opponent's facial expression, so that the other boy's gaze fleetingly becomes his gaze, the boy's smile becomes his smile, the boy's frown becomes his own. In this posture, the winner finds himself thinking and feeling in the same way as his rival. His success hinges entirely on the precision of his mime.

In a sense, we are always evaluating and predicting the other, though we may not heed the act. Often the people we scrutinize hardest are those we cherish the most. In love there is constant contemplation and the most intense desire to understand the object of our affections. There exists melancholy, too, as we grow to appreciate just how little we can ever truly know for sure. Our ignorance is painful. And yet we persevere. Humbly, patiently, we assiduously observe till at last we identify ourselves in some way with the other. Anticipation becomes an act of love.

I have spent years learning how to evaluate my mother's various tics and gestures. These days, I can read her body language with fair fluency. But still the same questions return over and over again. What, I often wonder, does that smile of hers say?

We meet in a posh restaurant in central London. A woman's face, old enough to be thought young-looking, smiles from a table at the back when I come through the door. I kiss my mother on the cheek. Her ecstatic eyes follow the stiff-backed young waiters carrying platters of food and wine. What shall we eat? I know the place well and have already made up my mind. I confide my choices to my mother before heading to the bathroom. When I return, the menus are gone. My mother is fiddling with her napkin. The skin on her hands is worn tight, I notice, like the peel on overripe fruit. Her fingers twist the napkin while we talk.

Her rent has gone up again. Arabesques of graffiti continue to shout from wall to wall. Last week, an Albanian down the road set fire to his mattress; the wailing sirens made "a right old racket." And yet, my mother will not hear of moving away. She insists on sticking close to where her children were all born and raised. I am aware that any new plea will fall once more on deaf ears, and have no choice but to let the matter go.

"Midday," I answer when she asks about my flight tomorrow

out of Heathrow. Tokyo, I tell her, is nine hours ahead of her time. The trip will include my first lecture in the Far East. My mother feigns curiosity. She has never held a passport, knows nothing of the world beyond her shores. All of a sudden, she shakes with silent laughter. Some word, its sound or the mental image elicited by the word, has tickled her. Like the boy in Poe's tale, I reciprocate and try to laugh along. But I do not understand. And then, just as quickly as it came, the laughter leaves her. With the napkin, she dabs the wet corners of her eyes.

I think back to the menu. For each course there are just a few options. A good opportunity to put my imaginary mother to the test. But will the model and the mother agree? Recalling the trios of starters, mains, and desserts, I assign to each a probability. I assess not only each of the individual courses but also the potential combinations across them. For example, two-thirds of the mains have meat: for this reason I downgrade the odds of my mother selecting pâté as a starter (unless, that is, she were to opt for a second course of fish). Assuming she begins with the good intentions of a salad, I plump for the caramel cake as her choice of dessert.

My imaginary mother decides to steer a middle route between the pâté and the salad. When the waiter looms, his voice announces the roast vegetable soup. I inhale the satisfying sweetness as it passes my chair.

Now the beef rises even higher up my mental shortlist. But when the table has been cleared, and the waiter next comes, I find my stare returned by the glassy eye of a baked cod. Bits of the fluffy flesh scatter around her plate and down her top as she eats.

Finally, we arrive at dessert. There is no room for doubt in my mind. My model mother's fondness for chocolate has been corroborated many times before. But not today. My actual mother finishes her meal with a bowl of exotic fruit.

Outside the restaurant, she takes my arm in hers. She wants to show me the street on which she grew up. She grips my arm tight as we take the short walk side by side. Of her life before me I know next to nothing, only one or two tidbits, gathered here and there. I ask whether it is true that before starting a family, she worked as a secretary. "Leave off," she laughs. "I typed addresses onto envelopes." She still has postal codes on the brain.

"Name a town," she says suddenly.

Bethlehem, I think. "St. Ives," I say.

"St. Ives," she repeats and it sounds twice as long when she says it. "TR26."

We turn onto the street, a stone's throw from Westminster Palace. My grandfather worked for the local brewery here, delivering the beer by horse and cart. The block of flats was home to several of the workers' families. A single washroom, with enamel bath, did for everyone. We cannot see inside, however. The building has become a base for the homeless; some of the windows are boarded up. Before we leave, I pull out my camera and take a photo.

I think of the girl who would grow up to become my mother. Who was the future woman that she imagined for herself? Did she dream of having a loving partner, a big house, and children who always smiled at her? In her mind's eye, would she be well educated and well traveled, always generous, patient, and kind? Did she imagine that every cherished moment would be remembered forever; every grievance instantly forgotten?

And thinking of this girl I feel at once immensely happy and immensely sad. I feel as I feel when I think of myself.

Twenty-One

TALKING CHESS

To win at chess is simple: victory belongs to the player who makes the next-to-last mistake.

Whoever it was that first came up with this line spoke much truth. The strongest players operate neither like machines nor angels; their superiority lies in accomplishing a better class of error.

A winning mistake would presumably owe nothing to sloppiness, incuriosity, or a yellow belly. It would be far closer to the lucky slip of an artist's brush or writer's pen, one that suddenly infuses a picture or page with unforeseen possibil ity. I am thinking of the (perhaps apocryphal) story of the painter who, hot-tempered at his many failed attempts to perfect a detail of his portrait, tossed his sponge with disgust at the easel and thereby achieved just the desired effect. Or that of the book printer who inadvertently won Herman Melville lavish accolades for the phrase "soiled fish" (in place of the intended "coiled fish" in reference to an eel).

I will not risk stretching this argument any further. Creativity is obviously so much more than an unexpected gesture here and there. But talent in chess bears this similarity with other creative pursuits, in tolerating error, as if the grandmaster—like the great artist—is the one who truly explores possibility's outer limits. Or, as a character in Joseph Conrad's *Lord Jim* intones, "To the destructive element submit yourself and with the exertions of your hands and feet in the water make the deep, deep sea keep you up."

Chess is a perfect arena for just such an exerted exploration of the possible. Its checkered sea is very deep indeed. The mathematics behind the game's complexity are staggering. An initial move by each player will create one of four hundred legal positions. A second move by each: seventy-two thousand. There are some nine million possible setups after the players' third move; 288 billion after their fourth. Back in 1950, the mathematician Claude Shannon calculated the possible number of forty-move games, a figure henceforth known as the "Shannon Number." He estimated as about thirty the potential number of viable moves at every turn. In this way he arrived at a total number (1 followed by 120 zeroes) that easily exceeds the number of atoms in the observable universe.

For all its immensity, chess is a finite game. It is therefore at least conceivable that a machine might one day be programed with the knowledge, deep down in its nodes, of every possible sequence of moves for every possible game. No combination, however ingenious, would ever surprise it; every board position would be as familiar as a face. Like checkers, which was solved by computer scientists in Canada in 2007, we would finally discover how chess, played perfectly by both sides, ends.

This perfect game of chess—the immaculate order and configuration of its moves, the exquisite ballet of its pieces in their precisely timed roles—is imprinted on the imagination of every player. In his innermost being, every player carries some notion of the divine game. For one, it begins with the two-square forward march of the white king's pawn, to which the L-shaped spring of black's queenside knight replies. Six moves later, the white queen arrives on a 4—a side square—only to be promptly rebuffed by black's bishop. No, no, says a second player: the game begins with a white knight—either one—to which black responds in kind. The central pawns advance in pairs. But another player disagrees, citing the fact that a piece would need to be sacrificed by white after eleven moves—the exchange of his queen for a rook. For still others the white pawns creep up the far sides of the board like ivy, or the black king crouches incessantly behind his queen, or all four bishops dance the diagonals until precisely half of all the original pieces remain.

Who, then, prevails in this Platonic ideal of chess? Each player confesses a secret faith. A triumph for white in forty-three moves, or in forty-one if black's sixth move disturbs a pawn. Or else a win for black, after a marathon two hundred and twenty-seven moves, the surviving white piece finally confiscated by the king. But such are the visions of romantics; a large majority seem resigned to the probability of a draw.

A small number of enthusiasts (a handful of masters, too) claim to have adjudicated the question in favor of one or the other side, outlining a "system" for the player to follow. Understandably these systems have been subjected to many critiques. In his books and articles, Weaver Adams, who won the U.S. Open Championship in 1948, argued that with a first-move king's pawn advance "white ought to win." In his own matches, however, Adams seemed to have perversely better luck with black. A certain Dr. Hans Berliner agrees with Adams concerning white's predestined triumph, but differs in his choice of the clinching initial move. According to Berliner, white must employ the queen's pawn instead.

Within our lifetimes a definitive solution, using the fastest computers, might indeed emerge. But there is still a very, very long way to go. So far algorithms have resolved every lawful chess position containing at most six pieces (including the two kings). The compiled data has yielded more than a few surprises. Numerous endgame positions long considered mere draws, we know now beyond all shadow of a doubt, can in fact produce a winner. Among the most recent analysis, which has moved on to the study of endgames of seven pieces, researchers have hit upon an astonishing forced win for white—provided that the winner's concentration could somehow hold out over five hundred and seventeen flawless moves!

In the end, maybe no solution of chess exists, or at least none that we could ever extract in the time allotted to our universe. Any complete solution might even reveal itself to be simply beyond the pale of our imagination: knights that forage for pawns for no apparent reason, bishops that take turns occupying consecutive squares, or rooks that slide up and down, and left and right, ninety-nine times in a row.

But of course chess would not be chess without its mystery, or its players' mistakes. Men, like chessmen, are made from crooked timber. With his mistakes, the beginner (what players call a "patzer") immediately blows his cover. He brings his queen out too quickly, exchanges too many pieces too soon, moves his pawns in such a way that their formation finishes by resembling Swiss cheese. But the problem is more one of quantity than quality. The patzer loses not because he makes too many mistakes, but because he makes too few—a mere handful of classic blunders. He does not last long enough to make more! Traps abound, and he duly falls into one or another. Cannier, moderate-strength players (such as club champions) make more mistakes than beginners. Sidestepping the early pitfalls, they grant themselves far greater scope in which to stumble.

Stronger play demands more than the avoidance of blunders. The player has to learn to make his own mistakes, which is far more difficult than it sounds. He has to stop mimicking the moves he reads in books and magazine columns, and which he does not really understand: even the best moves can turn bad when played on the wrong square or at the wrong instant. He has to root out his most cherished errors, the kind that he plays with the frequency of a tic. In short, he has to clear his head and think and feel and suffer for himself. Only then can he hope to obtain the slightest grip upon the game.

All of this amounts to nothing less than that nebulous

attribute we call personality. It is the indescribable quality that seems to bring the pieces on the board to life. As with the brushstrokes of a gifted painter, we can identify a strong player from the moves, including the mistakes, which he makes. The observer traces in the movements of the pieces the movement of the player's thought. What we call his mistakes are also the expressions of a deep and personal understanding of such-and-such position, which is—like every human understanding—imperfect. Out of this personal understanding comes both his biggest howlers and his finest moves.

No master of the game, I would suggest, had more personality than the Soviet champion "the Magician of Riga," Mikhail Tal. Not a few of his games have gained the status of the masterpiece. At its best, Tal's play betrayed a courage that bordered on insouciance. He was always in the thick of the action, inviting complications. Of this proclivity for problems, he once remarked, "You must take your opponent into a deep dark forest where 2 + 2 = 5, and the path leading out is only wide enough for one."

In the depths of this forest even he sometimes lost his way. During one simultaneous exhibition against a score of Americans, the grandmaster took the fight to a plucky and talented twelve-year-old. At a crucial moment Tal surrendered his queen in exchange for the initiative, but the sacrifice eventually proved unsound and he went on to lose. With a shrug of his shoulders, the former World Champion gamely shook the boy's hand, before continuing to ply his trade among the other boards.

Tal played instinctively. Faced with the game's intractable complexity, he always followed his nose. He would feel his way around the squares of the board, for feeling is also a kind of thinking. In his autobiography, there is a small but marvelous anecdote of this intuitive approach. Tal describes a contest with the grandmaster Vasiukov at the Championship of the USSR. By dint of adventurous play, the two men had reached a highly entangled position. Tal says he hesitated a long time over his next move. His path to victory, he sensed, began with the sacrifice of his knight, but the immense number of possible variations unnerved him. Head in his hands, he meditated on one after another without result. Mental chaos ensued. All of a sudden, from out of nowhere, an amusing couplet by the children's poet Chukovsky broke the surface in his mind. *"Oh, what a difficult job it was. To drag out of the marsh the hippopotamus."*

Tal was at a loss to know by what process of connection his mind had suggested the poem's lines. But now the thought gripped him: how exactly would a man pull such an animal from a marsh? As the spectators and journalists looked on, the grandmaster contemplated any number of hippopotamus-rescue methods: jacks, levers, helicopters, and "even a rope ladder." Here, as well, his calculations came to no avail. "Well, just let it drown," he said to himself at last, in a fit of ill temper. Tal's head cleared immediately and he resolved to trust his instincts and play. The next morning the newspaper report read, "Mikhail Tal, after carefully thinking over the position for forty minutes, made an accurately calculated piece sacrifice."

Before we leave Tal, one word about his chess education. The young Mikhail's development proved vertiginous. He first learned to play at the age of eight, watching the games of patients at a hospital where his father worked. The boy did not shine, far from it. His youthful style was simply one more against which the older players gained their points. Only from the age of twelve did he apply himself seriously. A local master took him under his wing. Within two years the teen had qualified for his national championship; a year later he finished ahead of his trainer. The next year, aged sixteen, he won his country's chess crown and the title of master.

Such rapid acquisition is reminiscent of the facility with which we learn our mother tongue. Only four years separated the beginner Tal from his first national championship; only four years—as a rule—separate the baby from fluent speech. In both cases, an adult's guiding hand makes all the difference. Left to his own devices, neither the baby nor the beginner can hope to make much progress. Linguists say that a child learns his language by exposure to highly structured input; his parents address him more slowly, more questioningly, in short sentences that verge on curtness. The chess player, in a similar fashion, learns best from a coach's counsel; he is shown many patterns and combinations of moves that characterize expert play.

Wittgenstein observed that language, like chess, is a game governed by rules. Knowing how to use a word, he said, is like knowing how to move a chess piece. From a small number of these initial rules, immense complexity is spawned. Gossip on a street corner can rival in complexity any game of chess. This is because the number of concatenations of words that form meaningful sentences approaches infinity. Speakers (and writers) are always coming up with original sentences, like chess masters with their novel moves. And like any player worthy of winning, the speaker (and again the writer, to some extent) anticipates the other's response. He modifies his speech to accord with his anticipation. Not only does he know what he can say, he knows what he should say, and—perhaps even more interestingly—what he should not. A skilled conversationalist has this knack for knowing which avenues to explore and which to avoid. Similarly, certain chess moves, in certain situations, while perfectly legal, are considered taboo. I have heard of a grandmaster who once described his opponent's early capture of a pawn with his bishop as being "low culture." The move boasted the advantage of quick material gain, but at the expense of the formation and coordination of his pieces. Such moves are rarely propitious to a good game.

I am reminded of a scene from the 1951 novel *The Master of Go* by Japanese author Yasunari Kawabata. The story's narrator, a passionate amateur of the game (Go is an ancient and highly strategic board game), plays on his pocket magnet set during a train ride home. Opposite him counters a tall American tourist, on whose knees the gold leaf board rests for the duration of the journey. The Japanese narrator prevails quickly, game after game. "It was as if I were throwing a large but badly balanced opponent in a wrestling match." Throughout, he notices a sort of thoughtlessness, a lack of personal investment, in the American's play. For the tourist, he imagines, playing Go was "like having an argument in a foreign language learned from grammar texts."

Taking this thought to its natural conclusion, Kawabata goes so far as to declare the game's subtleties as being inaccessible to foreigners. What he means by this, I think, is that a good game (whether of Go or chess), like a good conversation, requires a certain sensibility born of complete immersion. I am thinking of that attention to form which elevates a phrase or move beyond the merely functional. Even in translation, Kawabata's own elliptic writing style can often puzzle the non-Japanese reader. Many details, which to his countrymen appear straightforward, can pass entirely over our heads. Why, for instance, does the narrator in the novel mention that his Go board is decorated with gold leaf? Might it be a reference to illumination? Or victory? Should we read in it the suggestion of Go as an art? We can only hazard a guess as to its meaning.

Grandmasters, of course, are immersed in their game. Some drown, sunk by an insanity brought on by virtually infinite complexity. Most, however, find their own fullest and richest means of expression in the combinations of moves they deploy. These players do not so much think about chess, as think *in* chess, just as we think in language. I read an account of one master who recalls each day's events as though they were moves on a chessboard. He remembers going for an afternoon swim as, say, King's knight to F6, while a restaurant meal with his wife comes back to him as a descent four squares of the Queen's rook.

These associations appear to him as unremarkable as they are spontaneous.

We can also see this spontaneity, the spontaneity of speech, in what we call "blitz" or "rapid" chess. In its most popular form, both players have just one minute in which to dispense all their moves. Pieces slide frantically from square to square; agile hands swat over and over the button on their clock. At a rate of one move per second or thereabouts, whole games, upward of forty moves, are frequently accomplished. In spite of having no time to think, the players in these games often attain surprisingly high standards.

I don't wish to imply that playing chess is all instinct. Alacrity has its virtues (the absence of indecision for one), but they are not the virtues of the longer form. At its best, chess—like language—privileges reflection and careful thought. Looked at one way, a game of chess amounts to a long series of shifting problems, the most intriguing among them capable of making unique demands upon our imagination. As with a beautiful passage in a novel, we feel that we could happily spend almost any amount of time in its company.

Occasionally an amateur player will scour the back pages of certain newspapers (the kind whose pages end up wrapping glass vases, but never greasy fries) to savor and solve these chess problems. Illustrations of the board, its pieces frozen in print, compete for space with the week's crossword puzzle. "White to play and mate in three" or "Black to play and draw" its title will announce. Quite often the position displayed is at least half cleared of its starting pieces, and the game has attained its final moments. The amateur stares at the inky pieces and the smudged squares, waiting for a surge of inspiration. It is more or less the same experience as when we read some striking lines of a poem.

Many of these published chess positions have never known an actual board; they are the figment of some inventor's imagination. Among these inventors figures one "Vladimir Sirin," better known as the multilingual novelist and poet Vladimir Nabokov, for whom such compositions equated to the "poetry of chess" (his 1969 anthology, *Poems and Problems,* was illustrated by eighteen of his own efforts).

Those sixty-four squares, their peculiar geometry, fascinated the great wordsmith to the point of obsession. The right-angled triangle, for instance, traced by a king's vertical (or horizontal), and then diagonal, retreat across the board disobeys Pythagoras's famous theorem. Off the board, in the "real world," moving, say, three paces across and as many paces up (or down) would produce a greater diagonal distance (four and a quarter paces long) between the start and end points. The king's "triangle," on the other hand, has identical length (traversing the exact same number of squares) on every side. This optical illusion (expecting a diagonal route to require a greater number of moves than either a vertical or horizontal path) plays an essential role in the chess problem composer's craft.

Like his words, Nabokov's chess pieces relied on precise positions and combinations for their meaning. He saw to it that the pieces' values differed sharply from the amateur's expectations. The "value" of a position's queen, for example—while generally considered to be twice that of a rook, thrice the worth of a knight or bishop, and nine times as important as a pawn—might decrease to next to nothing if immobilized to a corner or back row by the careful setup of lesser pieces.

Numerous mathematicians also double up as inventors of chess problems. They create positions to solve questions like, what is the maximum number of possible checkmates in one move? Answer, forty-seven. Or what is the smallest number of bishops needed to occupy or attack every square? Answer, eight (the same as for the smallest number of rooks). Or else they compose whole games of a specific number of moves to reach a given position.

There is a final analogy between chess and language that I should mention. I began these pages reflecting on mistakes. I said that grandmasters make magisterial mistakes, founded on great creative intuition. Young children do, too. From ambient talk they get wind of words, but their minds then make of them what they will. As a matter of fact, the toddler's speech far exceeds the mimicry of adults; its features are distinct and demonstrate invention. For example, we have all heard a young child say something like "three mouses" or "I goed" instead of "mice" and "went," though no parent would utter such a thing. More inventive still were the kids who reportedly suggested that a "myth" was a female moth, or that benign is "what you will be after you be eight."

Perhaps they grew up to become grandmasters.

Twenty-Two

SELVES AND STATISTICS

For each one of us, nothing is so personal, so intimate, and so selfish as our death. From time immemorial, men have sought to predict its hour. They have scattered bird entrails, or elaborated bad dreams, or consulted the office of an oracle. When these prophecies were not plainly wrong, deluded, or hysterical, they often proved unhelpfully vague and misleading. According to one legend, a Scythian prince consulted a Greek oracle to discover how he would die. A *mus* (mouse), he was told, would be the agent of his demise. The prince took the forewarning to heart. He had his houses cleared of mice and even snubbed the company of any man called Mus. All the same, shortly thereafter, death came for him. The cause? He died from an infected arm muscle (the word *muscle,* in Greek, means "mouse").

The idea of death as a statistical phenomenon, amenable to calculation, occurred only toward the end of the seventeenth century, with the publication of the world's first mortality table. This new conception of death had gestated so long for a good reason: it upended completely how men and women understood their individual lives. *Society*—if and when this term existed at all—had always been held to be only a loose confederation of free souls. The precise business and destiny of each were an impenetrable mystery. Crowds were merely freaks, monsters with many heads and countless limbs. That a person might be in any way comparable to the crowd, that he might be deciphered by studying the behavior of his family, his village, his fellow countrymen, seemed both impossible and absurd.

And if the individual was beyond the pale of understanding, so, too, was his eventual extinction. No scholar, bitter experience told most people, could hope to get inside the Grim Reaper's skull. Death touched rosy-cheeked infants and decrepit widows alike, with no apparent rhyme or reason. An ancient man on his last legs might somehow survive another decade, while his grandson, beaming with youth and health, not live to see the following spring.

Stories took the place of science, their tellers repeating the same message over and over: life is full of surprises. Remember Old John, one tale would go, Old John who laughed so hard at his neighbor's joke that his heart gave out? The farmer's wife who was butted into her grave by the goat? The squire who caught cold sleeping in church?

It was in this atmosphere of ambivalence that Edmond Halley, who found lasting fame calculating a comet's orbit, published *An Estimate of the Degrees of the Mortality of Mankind* in 1693. Halley based his figures on the city of Breslaw, capital of the province of Silesia, "near the confines of Germany and Poland and very near the latitude of London," with a total population of 34,000. For five years running, monthly figures for every birth and death in the city had been collated: 6,193 births and 5,869 burials in all. Of the newborns, Halley discovered, 28 percent perished in their first year; only a little more than half lived to celebrate their sixth birthday. Most of these, however, would go on to have children of their own. "From this age the infants being arrived at some degree of firmness grow less and less mortal."

Of citizens between the ages of nine and twenty-five, the number of deaths each year equated to about 1 percent. This figure rose to 3 percent for those aged between twenty-six and fifty, jumping to 10 percent for those having reached the ripe age of seventy. "From thence the number of the living being grown very small, they gradually decline till there be none left to die."

Halley used this combined data to calculate "the differing degrees of mortality, or rather vitality in all ages." For example, to estimate the odds that a person aged twenty-five would not die within the next twelve months, he compared the city's twenty-five-year-olds (totaling 567) to its twenty-six-year-olds (560), and thereby concluded that an "average" twenty-five-year-old would have odds of 560 to 7, or 80 to 1, that he outlive the year.

What is the probability that a man of forty live a further seven years? Halley took the number of forty-seven-year-olds (377) and subtracted it from the number of forty-year-olds (445) to find the difference between the two ages (68). The odds that a forty-year-old become a forty-seven-year-old were therefore 377 divided by 68, or 5.5 to 1.

How many years might a man of thirty reasonably expect to have still ahead of him? To answer this question, Halley would first determine the number of thirty-year-olds (531), and then halve it (equivalent to considering as even odds that the person would die within the period). This halved figure (265), he then finds, equates to the number of citizens between fifty-seven and fifty-eight years of age. The "average" thirty-year-old could look forward to a further twenty-seven or twenty-eight years.

From his findings Halley drew a suitably pious conclusion.

> Unjustly we repine at the shortness of our lives, and think our selves wronged if we attain not old age; whereas it appears hereby, that the one half of those that are born are dead in seventeen years time . . . so that instead of murmuring at what we call an untimely death, we ought with patience and unconcern to submit to that dissolution which is the necessary condition of our perishable materials, and of our nice and frail structure and composition: And to account it as a blessing that we have survived, perhaps by many years, that period of life, whereat the one half of the whole race of mankind does not arrive.

It was a mortality table just like the one first drawn up by Halley that the American paleontologist Stephen Jay Gould would scrutinize at length, three centuries later, in the summer of 1982. It was the summer when the Equal Rights Amendment was voted down, when Italy beat West Germany in the final of the soccer World Cup, when a debt crisis broke loose across South America, and when Gould sat in his doctor's office and learned that he would shortly die. He was forty and had just been diagnosed with a rare and incurable form of cancer. Poring over books at Harvard's medical library, thick as hand-spans, subsequently taught him all that there was then to know about the condition and its survival rates. In summary: he had a median of eight months left to live.

Gould could not take seriously Halley's well-meaning words of advice. He would not submit readily to his body's dissolution; he wanted to survive. He thought of his wife, his two young sons, his high-flying career. A million other things went through his mind: the museum's Tyrannosaurus rex, all gigantic teeth and bulky bones, that he had seen as a small boy; his father, in his evening slippers, reading *Das Kapital;* the opening bars of Gilbert and Sullivan's *Mikado;* his Yankees season ticket; the Pepperidge Farm cookies in his office drawer. His office. What would his microscopes do, and his favorite rattan chair? Sit on the wrong shelf; stand in an unlit corner. Gather dust.

What did Gould do in such dire circumstances? What could he do? He did what virtually every recipient of bad news does: he hunted feverishly after even the slightest, even the faintest positive angle. He would not abandon hope. A *median* of eight months; that is what the statistics said. If half of all the patients with his cancer would die within eight months of their diagnosis, it meant that half would live beyond eight months. Some would live on for years. This thought comforted him. His mind tightened around it. Age? He was still young. Class? His family lived on the better side of town. Health? A little on the heavy side, but he had no other baggage. Attitude? He identified a strong will, even temper, and a clear purpose for living. His chances of landing in the latter half of the patients' prospects seemed to him great.

He would have only one future death, not thousands, and the median said next to nothing about it. This became his mantra. Friends and family asked him to explain. Averages talk about populations, not persons, he would reply. If I died a thousand times, approximately half of those deaths would occur within eight months. The other half would follow one by one: days, weeks, months, or years later. Who can tell where my single death, out of the thousand possible deaths, will fall?

The following months were tough and turbulent for Gould, full of boredom, pain, and exhaustion. Doctors radiated his body, pumped it with drugs, and put it under the knife. He hemorrhaged weight—in all, he lost a third of his 180 pounds. His hair embarrassed him by falling out. The lonely and tedious hours of treatment piled one upon the other, oppressing and enfeebling him. And yet he survived. His cancer went into remission. Two years later he was well enough to write a long article about his experience, "The Median Is Not the Message." A decade after its publication he was still going strong. "I am a member of a very small, very fortunate, and very select group—the first survivors of the previously incurable cancer," he wrote.

In March 2002, Gould—aged sixty—published his magnum opus, *The Structure of Evolutionary Theory,* 1,342 pages long. It was the seventeenth book to bear his name since his diagnosis twenty years before.

Two months later, his personal death finally arrived, the result of a second, unrelated cancer.

Knowing how to read the mortality table's numbers probably extended Gould's life by many years (he affirmed a likely link between an individual's state of mind and his immune system). Not knowing how to read them, on the other hand, cost one man and his family very dear.

André-François Raffray's story is an extreme illustration—perhaps the most extreme—of confusing persons with percentages. Raffray was a long-serving and successful public notary, living and working in the southern French city of Arles. Among his clients was a ninety-year-old widow, without heirs, whose name was Jeanne Calment. One day, in 1965, Raffray agreed to purchase the widow's house under a scheme the French call *rente viagère:* in return for paying Calment a monthly sum of 2,500 francs, he would own the property upon her death.

Raffray must have imagined that he had struck an excellent deal. The widow's house was worth close to half a million francs. Assuming that she lived another three years—the current life expectancy for an average ninety-year-old French woman—he would shell out fewer than 100,000 francs in all. Over 20 percent of nonagenarians died before their next birthday. The statistics, he thought, were on his side. "Even if she makes it to ninety-four, or ninety-five, or ninety-six, I will still have the house in the end for only a fraction of its value. And what if she goes on and on, to ninety-seven, or ninety-eight, or — God forbid — to one hundred? But how many people ever live to one hundred? Not even one person in a thousand! To think that she might go on another ten years! I cannot imagine it. But what do I care? Let her keel over at one hundred: I will still make a tidy profit." Such was probably his line of thought.

It was a mistaken assumption that the very old are all more or less the same. The notary's acquaintance with Madame Calment was slight. Her snow-white hair, her bird-like frame, her papery skin, he mistook for frailty. He saw these features and thought at once of every elderly person he had ever met. Faces, bodies, lives, blended together in his mind. What did these people, averaged, have in common? Sickness, sadness, shortness of breath.

But before Madame Calment was old she had been young, and she had ridden a bicycle through the cobbled streets of Paris, and stretched after fluffy tennis balls, and eaten scores of fruit and salads slick with olive oil. Early marriage to a wealthy merchant had freed her hands for tinkling piano keys and theater applause. No illness had ever troubled her.

Little had changed for her since moving to the warm and sunny south. She had taken all her favorite things with her, except for the buried husband. But she had long grown comfortable with her solitude. She was not afraid of silence, of hearing the rhythm of her heart. Nor did she worry about her looks: makeup, she had learned, could not resist her frequent tears of laughter. At age eighty-five, she wore silly plump padding for her first fencing lessons without giving it a second's thought. She still loved walking in the open air. Regular sips of port wine and snacks of chocolate brightened her day.

Careful study of the mortality table had taught the notary with what frequency previous nonagenarians had died and after how many years, but he had not thought about the future. It was a fact, for example, that France counted no more than a few hundred centenarians in 1965. But that was then, and the widow would reach her hundredth birthday, if she reached it, in ten years' time. How many centenarians would France have in 1975? In 1980? In 1990? That is the kind of question that the notary forgot to ask. Around the world, medicine and technology were rapidly improving. Once significant causes of death—from flu, vitamin deficiency, or high blood pressure—were dwindling. Within a generation, the number of centenarians in France would increase twentyfold.

And what of the table's statistics? They deserve a closer look. For one thing, the data available for the very old was necessarily sparse and unreliable. Too few before the generation of Madame Calment had lived long enough to give being ninety years old a proper try. Statisticians knew next to nothing of a nonagenarian's medical needs, eating habits, daily routines, and much else besides. Guesswork had to fill in the gaps.

Life expectancy: three years, announced the table. But let us see what this actually means. If there were ten thousand ninety-year-olds in 1965, there would be around five thousand still living in 1968. The life expectancy of these ninety-three-year-olds would not be zero. What would it be? (This was another question that the notary did not think to ask.) Almost three years. Live three years in your nineties and the chances are fair you will live three years more. And if there were five thousand ninety-three-year-olds in 1968, there would be around two thousand surviving in 1971. These ninety-six-year-olds could expect, on average, to live a further two years. In 1973, about a thousand—10 percent of the original number of ninety-year-olds—would still be alive. Close to half of these in turn would live long enough to reach their hundredth birthday.

Madame Calment, in February 1975, numbered among them. One hundred years old, but still in fine shape, on her feet every day. She left the dying to others. At the age of 105, her notary's monthly checks amounted to the full value of her house. But still he had to keep on paying.

Five more years passed and Madame Calment begrudgingly moved into a nursing home. At 110, her age exceeded the maximum figure of most mortality tables. At 113 she became the oldest person in the world: *la doyenne de l'humanité*.

Even then, she did not die. Her eyes dimmed, her joints stiffened, but she remained in good spirits. Statisticians who had long estimated a maximal human life span of between 110 and 115 years were proven wrong. Madame Calment became the first person in history to celebrate her one-hundred-and-sixteenth birthday, her one-hundred-and-seventeenth birthday, her one-hundred-and-eighteenth birthday, her one-hundred-and-nineteenth birthday.

In 1995, "Jeanne," as people now called her, attained 120 years of age. The nursing home, which until then had received few visitors, suddenly swelled with reporters from around the world. Tiny, desiccated, she sat before a bank of cameras. She does not smile, complained one of the photographers. Ask her if she wants to go on living, demanded a journalist. The nursing home director cupped her hand around the old woman's right ear, as though playing an absurd game of Chinese whispers. "The monsieur would like to know if you want to live a little longer?" she shouted into the ear.

Yes.

Her notary was absent that day. Long retired, nearing eighty, he was not well enough to attend. Later that year, he died. He had never set foot inside Madame Calment's house. Under the terms of the thirty-year-old contract, his widow continued to pay. By the time Madame Calment eventually died two years later, aged 122, the notary and his family had paid nearly a million francs: many times the anticipated payout and twice what the property had been worth when they took on the arrangement.

From where, we might well ask ourselves, did this strange idea of an average cancer patient or an average nonagenarian originate? *A Treatise on Man and the Development of His Faculties* introduced *l'homme moyen* ("the average man") to the nineteenth-century world. "If an individual at any given epoch of society possessed all the qualities of the average man, he would represent all that is great, good, or beautiful."

The author—a Belgian mathematician called Alphonse Quetelet—imagined how Nature created men. He pictured Nature as a bowman who aimed continuously for the statistical center. The bull's-eye would be a perfectly median man (or woman): a rational, temperate person free from any excess or deficiency. Most individuals, according to this view, were Nature's errant arrows. They straggled, nearer or farther, around the mean. The taller-than-average flaunted the shorter man's missing inches; the drunkard lacked his part of the teetotaler's (excessive) self-control; the hairy man wore the bald man's follicles.

Quetelet therefore urged his readers to take a wide view of humanity. Every person, after all, was "but a fraction of mankind." Instead of interesting themselves with men and women, scientists should study the population as a whole. "The greater the number of individuals observed, the more do individual peculiarities, whether physical or moral, become effaced, and allow the general facts to predominate, by which society exists and is preserved."

So the mathematician averages data collected from a thousand or ten thousand households and learns, for example, that the "normal" male height is five feet seven inches, that the "normal" time spent reading the newspaper is twelve minutes, that the "normal" diet consists of eggs, potatoes, and meat. A height of five feet five inches, or six feet two inches, is aberrant; spending five minutes, or thirty, with the newspaper is abnormal; too much fish or too few eggs on a plate is deviant.

Drawing from this data, the mathematician observes

regularities: most men are two inches too tall or too squat, most readers linger three minutes too long (or skim three minutes too short) over their newspaper, most housewives cook six potatoes per week too few or too many.

But it was not only physical or mental traits that a mathematician might average. Morality, too, could offer itself up to calculation. An analysis of police statistics would reveal the defining characteristics of the "average" criminal, "even for such crimes that seem to escape all human foresight like murder since they are committed in general without motive and in circumstances, apparently, the most fortuitous." According to Quetelet's figures, the "typical" murderer would be male, in his twenties, literate and a white-collar worker. He would have alcohol on his breath, and be wearing the light clothes of summer. Probability would put a pistol (rather than a knife, or a bat, or a vial of poison) in his hand.

This idea spread quickly. It proved popular with scientists and the general public alike. Most frequently it was reinforced neither by logic nor analysis, but by simple prejudice. People laughed, and sneered, and denounced, and ridiculed variously identified "types" of the average man. But worst of all, they believed that such beings really did exist.

Images propagated the idea with particular efficacy; as the proverb says: a picture is worth a thousand words. One newspaper caricature was enough to do the work of Quetelet's wordy disclaimers (his original book had run to several hundred pages). If the caricature depicted an Irishman — all Irishmen — with a misshapen jaw, feathered cap, protruding teeth, it was plain enough that the "average Irishman" bore more than a passing resemblance. If it showed a filthy, vulgar, gin-swilling beggar, it confirmed how many readers imagined the "average poor."

Photography—then still a young and innovative technique—was similarly enlisted. Mug shots of eight different individuals were superimposed to reveal the blurry face of the "average criminal." Nine photos of consumptive patients were merged to produce a portrait of tuberculosis. Images from six medals depicting Alexander the Great were morphed to reveal the ancient king's probable features. "It is now proposed," announced one magazine, "to get a clear idea of Nebuchadnezzar from the various stone and brick slabs upon which his face is graven."

But if photography served to popularize the idea of "average men," it also produced an alternative way of looking at ourselves. Identikits (facial composites), created at the end of the nineteenth century, helped to divert some of the focus back from the typical toward the individual. Photos that magnified and emphasized all manner of different facial features now replaced the phantom faces typifying this kind of man, or that sort of person. Instead of a single abstract representative of this or that contrived category, the images highlighted a wealth of actual noses, foreheads, wrinkles, ears, chins, eyelids, lips, and mouths.

Take the nose, for example. Plainly, nobody has a "nose"—a person has a low nose or a high nose, a wide nose or a narrow nose, a hooked nose or a straight nose, a long nose or a snub nose. Perhaps the tip is bulbous, the nostrils dilated. And what about the chin? Big or small? Flat or bumpy? Does it retreat toward the throat, or jut out proudly? Does it form a square, or slope down into a point?

A new picture of the person emerges. Yes, of course, commonalities remain. His name equally belongs to other people, his nose to other faces. We are all made of the same blood and bones. But take a closer look. See the proportions, the interplay between all the various parts? Every combination, like a mosaic, is unique. He has his father's eyes, and his mother's curls, and his uncle's lopsided smile. Together they create something, and someone, new. Someone who will look through those eyes in his own way, who will wear that hair according to his own style, who will deploy that smile for his own reasons. Talk to that person. Watch the skein of laughter lines that diagram his face. And how his eyes glisten, or darken, at the sound of certain words. He is simply being himself.

Quetelet (and many others after him) believed that the essence of human nature could be found in the average, but he was mistaken. The essence of human nature is its endless variety. As Stephen Jay Gould would later remark, "All evolutionary biologists know that variation itself is nature's only irreducible essence. Variation is the hard reality, not a set of imperfect measures for a central tendency. Means and medians are the abstractions."

Twenty-Three

THE CATARACT OF TIME

If, as is often said, lifetimes flow like a river, they begin with a trickle and culminate in a cataract. Heraclitus, the ancient Greek philosopher, put it well when he said, "Time is a game played beautifully by children." Perhaps this is the root of nostalgia: less the desire to return to our early years than to the more capacious experience of time that we inhabited as children.

Time. You know how it goes. After the age of thirty, I found, the days begin to run away from us. We struggle to keep up. That is when the nostalgic impulse awakens, burgeons, pesters. Last year I moved to Paris. Something about being back in a big city after so many years meant I could not help thinking, more and more, about the old London neighborhood of my youth. I had reached that age when the past becomes so big and so deep that your mind finds itself increasingly drawn there. It is like living on a fragile coast, by an imposing sea whose smells and sounds gradually overwhelm your senses.

So I decided to return. There was salt water to cross, and buses that jolted and trains that bored, but none of those things mattered. I simply had to go and see the place again after all this time.

My younger siblings, when I spoke to them about the plan, chorused dissuasion. "There is nothing there," they said, perplexed. Evidently, they had all moved on. "Why go back?"

I tried to explain. I opened my mouth, felt the breath on my tongue form plausible shapes. But my heart was not in it and each of my made-up reasons fell flat. I decided not to argue. I booked the tickets, packed my bags, and left.

That I could not explain my desire to return to my old London home did not disturb me. On the contrary, its surprising strength convinced. Reassured, even. Gazing out from the wobbly train, I tried to remember when I had last set foot there. My reflection in the window wore an expression of thought. Tall trees and green hills ran past. I looked away, cracked open a book, and stared at the pages till the words seemed to gel into a single inky mass.

Five years. Already? Where had they gone? So many things that I had accomplished, people that I had met, places that I had seen, now looking back, seemed to have taken hardly any time at all. And yet how difficult, how exhausting, how important each event had struck me in the moment! And how impossibly distant, a lifetime away, these hours in the train out of Paris would have appeared to me back then.

It was a fast service to London, without delays. Arriving in the capital, I felt more like a suburban commuter than an international traveler. I changed trains and rattled out from the center toward the familiar periphery, my excitement building. Gradually the carriage emptied of its suits; a different class of passenger took their place. "We must be getting close," I thought, straining forward, oblivious to my wristwatch. Near the end of the line, I gathered my things—my bearings too—and stepped out. The platform was covered in litter and broken glass, but for an instant, at least, it felt unambiguously good to be back.

Time is more than an attitude or a frame of mind. It is about more than seeing the hourglass as half empty or half full. More than ever in this age, let us call it the computer age, a lifetime has become a discrete and eminently measurable quality. To date, to believe the surveys in newspapers, I have spent some one hundred thousand minutes standing in a line, and five hundred hours making tea. I have spent a year's worth of waking days on the hunt for lost things. This year, I knew, contained my twelve-thousand-and-twelfth day and night. That number equates to over a quarter of a million hours, seventeen and a quarter million minutes. Counting one number for every second since my birth, I had recently made it into the billionaire's club.

We occasionally liken time to money, as something to be spent wisely, but it is not money. No refunds are possible for days ill spent; no bank exists to take savings. We cannot apportion our time like money, since we live always in ignorance of when the former will end. How to plan when a person can never know if he will see tomorrow, or survive to such an age that his eyes turn coal-black with blindness?

Perhaps it would be better to talk about time in the

manner of certain tribes. Strangers to clocks, they pace their days according to nature. Native Americans traditionally planted corn "when the leaf of the white oak was the size of a mouse's ear." Equinoxes and solstices scheduled their rituals. As for language, the Sioux have no word for "late" or "waiting."

In Australia, the Aborigines believe that time, place, and people are one. A glance at a tree or a face suffices to know the hour and the day. Their discrimination of the seasons is precise, depending on such factors as plant life and changes in the wind: the Eastern Gunwinggu, for example, speak of six seasons—three "dry" and three "wet"—where non-Aborigines see only one of each.

For these and other tribes, time is the product of one's actions. It appears when they sing a song, climb a mountain, or smoke a pipe, and vanishes when they sleep. They do not think of time as something pervasive, like the air. Seconds, minutes, and hours—these are all things that we do. In place of these terms, they speak of a "time of harvesting" or a "river fish time." Ask an African herdsman how long such-and-such task might take and he replies, "Cow milking time," meaning the time it takes to milk a cow. What is an hour to such a man? Perhaps the time it takes to milk ten cows.

We can put it another way: 1 hour = 10 milkings. My equivalent would be 1 hour = 10 tea-makings. Let us call it "tea time." A short walk that lasts eighteen minutes equates to three milkings or makings-of-tea; a two-minute commercial break amounts to one-third of a cup of tea. Between the opening and closing whistles of a soccer referee, time enough would pass to milk fifteen cows, or make fifteen cuppas.

I do not mean by this digression to suggest that approximations necessarily trump exactitude. It is not at all my intention to run clocks down. But the particular words and images our respective cultures deploy shape the way in which we experience time. I said just now that time is not money; we might say instead that it is closer to the spending of money. According to the tribesmen's way of thinking, it is what happens when, for example, we enter a marketplace. This emphasis on activity in how we think about time strikes me as being very healthy. When I hear someone complain about all the hours or weekends he has to fill, I stop and think that it is a mistake to speak of days as we would speak of holes. One hole is much the same as any other, whereas every day is different. In this, it is more like dough that we can sculpt into infinitely varying shapes.

On the journey back to my childhood home, I paused outside the train station, then made my way north toward the high street. The buildings were more or less the same as in my recollection: the same squat walls, tattooed with graffiti; the same "50 percent off" signs in shop windows; the same boys and girls, their busy fingers unwrapping candy. No bravado in the architecture, no color or charm. Along the sidewalks, no bustle either—either too early or too late for shoppers. Few cars animated the road. I walked mechanically, turning here and there, smelling the sugar of freshly laid tarmac on Waterbeach Road.

I landed finally on my old street. I took it all in. On the left stood metal railings and distantly behind them the classroom buildings of my former elementary school, factory-long. To the right, a chain of brick houses, close set. Their thin walls, I recall, made bad neighbors. Down the road, I spotted a small man in the distance. The man grew bigger with every step. He was wearing a blue-and-red jersey, but he did not look like a soccer player. The tightness of the shirt pronounced a sizable paunch. His dark hair was cut penitentiary short. His breathing rasped as he passed me by. And then he was gone.

I was surprised by how little had been altered. Painted house number signs, wooden gates, hedgerows all long forgotten, I recognize instantly. And yet it all seems so different from my kid days. Something has shifted out of sync, something I try to put my finger on. In frustration, I walk up and down the street until my legs tire. Only as I ready myself for the ride back does it hit me. What has changed here is time.

In his 1890 classic work, *The Principles of Psychology,* the American philosopher William James noted, "The same space of time seems shorter as we grow older — that is, the days, the months, and the years do so; whether the hours do so is doubtful, and the minutes and seconds to all appearance remain about the same."

James goes on to cite a mathematical explanation for this phenomenon, by a contemporary French professor. According to this professor, Paul Janet, our experience of time is proportional to our age. For a ten-year-old child, one year represents one-tenth of his existence; whereas for a man of fifty, the same year equates only to one-fiftieth (2 percent). The older man's year will thus seem to elapse five times faster than the child's; the child's, five times slower than the man's.

What matters, then, is the relationship between one sequence of years and another sequence. The interval spanning the ages of thirty-two and sixty-four will seem to the individual of similar duration to that experienced between the ages of sixteen and thirty-two, and to the interval between the ages of eight and sixteen, and to that from the age of four to eight, each having the same ratio. For the same reason, all the years from the age of sixty-four to one hundred and twenty-eight (assuming such an age were ever attainable) would seem to us to occupy no more of our feeling, thought, pain, fear, joy, and wonder than that big bang epoch between our second and fourth year.

More recently, from T. L. Freeman, we have a formula using Janet's insight that yields the individual's "effective age." Freeman's calculations suggest that we experience a quarter of our entire lifetime by age two, over half by age ten, and more than three-quarters by our thirtieth birthday. At only about the chronological midway point, a forty-year-old will experience his remaining time as seemingly but one-sixth of what has gone before. For a sixty-year-old, the future will seem to last merely one-sixteenth the duration of his past.

Are all our attempts to look back, to relive some bygone period, in vain? We can never walk down the same street twice. Those streets of my youth belong to another time, which is no longer my own. Except, that is, when I dream.

Fast asleep, I become a visitor there. I see a schoolgirl at the edge of the hopscotch grid, contemplating her throw. A man, atop a ladder, is washing his windows. His free hand glides rhythmically upon the glass. On the pavement, a neighbor's tabby squirms in the sunshine: stretching, and stretching his paws. The grunts and sighs of passing traffic fill my ears. I see my grandfather, alive, standing with his cane at the gate, as though keeping guard over my father's vegetable patch. I stop and watch my father. Sleeves hoisted to his elbows, he picks beans, sows herbs, and counts cucumbers. I watch without hurry, without a care in the world. Time is dilated; there is no time.

Our bodies keep time a great deal better than our brains. Hair and nails grow at a predictable rate. An intake of breath is never wasted; appetite hardly ever comes late or early. Think about animals. Ducks and geese need only follow their instinct for when to pull up sticks and migrate. I have read of oxen that carried their burden for precisely the same duration every day. No whip could persuade them to continue beyond it.

Asleep or awake, our bodies keep time a great deal better than our brains. We wear the tally of our years on our brows and cheeks. I doubt our bodies could ever lose their count. Like the ox, each knows intimately the moment when to stop.

Twenty-Four

HIGHER THAN HEAVEN

On January 22, 1886, Georg Cantor, who had discovered the existence of an infinite number of infinities, wrote a letter to Cardinal Johannes Franzen of the Vatican Council, defending his ideas against a possible charge of blasphemy. A devout believer, the mathematician considered himself a friend of the Church. God, he believed, had used his preoccupation with numbers to reveal a further aspect of His infinite nature. Fellow logicians had mostly sidestepped the young man's thinking; hardly anyone yet took seriously the outstanding insights that would make his name.

Before Cantor it had been impossible to speak mathematically about different sorts of infinity. All collections without a final object (the sequence of odd or even numbers, for example, or the primes) were simply conceived as being of equal size. Cantor proved that this was false. His papers were the first to demonstrate uncountable sets of numbers, that is to say, numerical sequences that even an infinitely long recitation could not exhaust. What is more, each uncountable set of numbers spawned another set of numbers that was even "bigger" than the last. Of the making of such sets, Cantor realized, there was no end.

The mathematician Leopold Kronecker, for whom "God created the integers [whole numbers], all else is the work of man," had no truck with Cantor's (infinite) tower of "smaller" and "bigger" infinities. He hounded his rival with violent words, called him a charlatan, a corrupter of youth. In the absence of his peers' understanding, Cantor turned at last for support to the Holy See.

The dialogue between theology and mathematics—varied, fitful, and singular—has a long history. Above all, infinity became the favorite topic. God is infinite, therefore mathematics is religion: a pathway to knowledge of the divine. This is what the Church fathers reasoned, and this is why the monks long ago proceeded where the mathematicians had feared to tread.

A thousand years before Cantor, in an Irish monastery, a man sat day after day at a table smelling of wicks and manuscripts. He spent years almost immobile, in deep and sustained contemplation, meditating on a perfect sphere that exists beyond space, universal and without limit. Of course, it is contradictory to think about a shape that has no border. The monk knew this. He knew that to think about infinity is to think in contradictions.

Minutes passed, hours passed. But what is a minute or an hour when compared to eternity? No time at all. A minute, an hour, a year, a thousand years, are all equally long or short in comparison. The light in the monk's cell would gradually disperse at the end of each long day; his mind might stutter, "I, I, I, I, I . . . ," but try as he would, Johannes Scotus Eriugena—John of Ireland—could not escape his senses and grasp the infinite.

According to Eriugena, God is not good, since He is beyond goodness; not great since He is beyond greatness; not wise since He is beyond wisdom. God, he writes, is more than God, more than time, *infinitas omnium infinitatum* (the infinity of all infinities), the beginning and end of all things, though He Himself had no beginning and will meet no end. Eriugena recalls the words of Job:

> Can you search out the deep things of God? Can you find out the limits of the Almighty? They are higher than heaven—what can you do? Deeper than Sheol—what can you know? Their measure is longer than the earth and broader than the sea. If He passes by, imprisons, and gathers to judgment, then who can hinder Him?

If God is infinite, Holy Scripture, being inspired by God, is held to exist outside the bonds of conventional time. Eriugena cites St. Augustine to affirm that the Bible often employs the past tense to express the future. Adam's life in Paradise "only began," occupying no real time at all, so that its depiction in Genesis "must refer rather to the life that would have been his if he had remained obedient."

Augustine's teachings contributed greatly to the Irish monk's thought, and that of the theologians who followed. In *The City of God,* Augustine insists that God knows every number to infinity and can count them all instantaneously. "If everything which is comprehended is defined or made finite by the comprehension of him who knows it, then all infinity is in some ineffable way made finite to God, for it is comprehensible by his knowledge."

Two centuries after Eriugena, in 1070, Anselm provided his famous "ontological proof" that God is that-than-which-nothing-greater-can-be-thought. If every number has its object, the object of infinity is God. Anselm became Archbishop of Canterbury; one of his successors, Thomas Bradwardine, in the fourteenth century, identifies the divine being with an infinite vacuum. The finite world is compared to a sponge in a boundless sea of space.

Infinity begets finitude, and thus cannot be grasped in finite terms. But how then to understand infinity in infinite terms? Alexander Neckham, a twelfth-century reviver of interest in Anselm's work, offered this problem a vivid image. For Neckham, God's immensity is such that even if one were to double the world in the next hour, and then triple it in the hour after that, then quadruple it in the following hour, and so on, still the world would be but a "quasi point" in comparison.

Such immensity inspires in the monks at once admiration and consternation: consternation, because an infinitely remote divine being would rule out the Incarnation. For the same reason, the believer would never see God in the Beatific Vision, and neither could he ever conform his will to the divine will. The vacuum is in fact a chasm, forever separating Mankind from its Creator.

The *De Veritate* of Thomas Aquinas, written between 1256 and 1259, offers a solution: "as the ruler is related to the city, so is the pilot to the ship." An infinitely powerful ruler bears no direct comparison to a humble captain, yet both possess a "likeness of proportions": a finite quantity equates to another finite quantity, in the same way that the infinite is equal to the infinite. In other words, "three is to six as five million is to ten million" bears a likeness to the proportion "God is to the angels as the infinite vacuum is to an eternal creation." Aquinas deploys the analogy throughout his work: as our finite understanding grasps finite things so does God's infinite understanding grasp infinite things; as our finite intellect is to what it knows, so is God's infinite intellect to the infinitely many things He knows; just as men distribute finite goods so does God distribute all the goods of the universe. Aquinas writes that the similarity between the infinite God and His finite creation constitutes a "community of analogy... The creature possesses no being except insofar as it descends from the first being, nor is it named a being except insofar as it imitates the first being."

Exasperated by critics he called "murmurers," Aquinas sought to settle a further point of contention. The Church taught that the world had a beginning in time. "The question still arises whether the world could have always existed." He penned these words in 1270, entitling them *De Aeternitate Mundi* (On the Eternal World). His argument was that if the world has always existed, the past regresses infinitely. The world's history must comprise an infinite sequence of past events. If there exists an infinite number of yesterdays, then an infinite number of tomorrows must also succeed. Time is infinitely past, and infinitely future, but never present. For how can any present moment arrive after infinitely many days?

Before this potentially unsettling line of reasoning, Aquinas remained unmoved and unimpressed. Halfhearted were his remonstrations. Any past event, like the present moment, is finite: therefore the duration between them is also finite, "for the present marks the end of the past."

And what about the succession of past events? Aquinas says the arguments for them can go either way. Perhaps God, in all His power, has created a world without end. If so, nothing obliged Him to populate it before Mankind.

A contemporary, Bonaventure, disagreed with Aquinas's equity. His blood thumped at the thought of an interminable past. "To posit that the world is eternal or eternally produced, while positing likewise that all things have been produced from nothing, is altogether opposed to the truth and reason." And what about the contradictions? For instance, if the world were eternal, tomorrow would be a day longer than infinity. But how can something be greater than the infinite?

In the fourteenth century, Henry of Harclay also faulted Aquinas for saying that an eternal world was possible, but from a point of view entirely opposed to Bonaventure's. For Harclay it was in fact probable, and every supposed contradiction dissolved on careful scrutiny. How can something be greater than the infinite? Look, said Harclay, at the infinite number of numbers: we can count from two, or from one hundred, and in both instances never reach a final number, though there are more numbers to count in the first infinity than there are in the second. He invoked Aquinas's proportions to defend the thesis of an infinite universe in which the infinitely many months occur twelve times more frequently than the infinitely many years.

To those who point out that an infinite past would have produced an infinite number of souls with infinite power like God, Harclay refutes the argument as follows: infinitely many souls would not constitute an infinite power. They would be not "any species of number, but a multitude of infinitely many numbers." Within this endless multitude, every possible number (59, 1,043,962, 999,999,999,999,999, 999,999,999,999,999 . . .) could be found, distinct and finite, each corresponding to a soul; save, that is, for an infinitieth number/soul since this would produce a contradiction: "there is not a number of infinite numbers, for then it would contain itself, which is impossible."

We trace to the same period, in the monk Gregory of Rimini's hand, the first definition of an infinite number as that which has parts equally great as the whole. Every twenty-third number, for example (we might just as well have taken every ninety-ninth number, or every third, or every five billionth), in the infinite succession of counting numbers (1, 2, 3, 4, 5, 6 . . .) produces a sequence as long—infinitely long—as all the counting numbers combined: match one with twenty-three, two with forty-six, three with sixty-nine, four with ninety-two, five with one hundred and fifteen, and so on, ad infinitum.

Gregory articulated his defining idea fully five centuries before Cantor. He taught for many years in Paris, at the Sorbonne, where his pupils called him *Lucerna splendens*. Perhaps in him they sensed, as future scholars would claim, the last great scholastic theologian to wrestle with the infinite.

John Murdoch, a historian of mathematics at Harvard University, remarked that Gregory's insight received hardly any notice from his peers or successors:

> Since the "equality" of an infinite whole with one or more of its parts is one of the most challenging, and as we now realize, most crucial aspects of the infinite, the failure to absorb and refine Gregory's contentions stopped other medieval thinkers short of the hitherto unprecedented comprehension of the mathematics of infinity which easily could have been theirs.

In his writings, Cantor described himself as a servant of God and the Church. His ideas had struck him with the force of revelation. It had been with God's help, he said, that he had worked day after day, alone, at his mathematics. But the mathematician was far from angelic; his humility sometimes slipped. To a friend, in 1896, Cantor confided an excess of pride. "From me, Christian Philosophy will be offered for the first time the true theory of the infinite."

Twenty-Five

THE ART OF MATH

I met a mathematician at a "conference of ideas" in Mexico at which we had both been invited to speak. He was from the United States, and like all the mathematicians that I have ever met in my travels he fell immediately to talking shop. Moving to a corner of the conference green room, he talked to me about the history of numbers in Cambodia. The concept of zero, he believed fervently, the familiar symbol of nothingness, hailed from there. He dreamed of trekking the kingdom's dirt tracks, in pursuit of any surviving trace. More than a millennium separated him from the decimal system's creation; the odds of turning up any new evidence were slim. But he did not mind.

He began to explain his current research in number theory, talking quickly with the compression of passion, and I listened intently and tried to understand. When I understood, I nodded, and when I did not understand, I nodded twice, as if to encourage him to move on. His words came fast and enthusiastic, opening up vistas that I could not quite see and mental regions into which I could not follow, but still I listened and nodded and enjoyed the experience very much. Occasionally I supplemented his ideas and observations with some of mine, which he received with the utmost hospitality. It always feels exciting to me, the camaraderie of conversation: no matter whether it involves words or numbers.

He had none of the strange tics or quirks of the mathematicians that we find in books or see in movies. From experience, I was not the least surprised. Middle-aged, he looked fit and slender, though with skin as pasty as a writer's. His shirt was open at the neck. His face wore many laughter lines. When our time was up, too soon, he patted his pockets and withdrew from one of them a small notebook in which he habitually jotted down his random thoughts and sudden illuminations. As he wrote out his contact information for me, I noticed the smallness and smoothness of his hands.

"Great meeting you." We promised to stay in touch.

It was still a pleasant surprise, coming down the next morning to the hotel restaurant for an early breakfast, to hear the mathematician's voice call me over to his family's table. I passed the assorted reporters munching their bowls of cereal, and various conference "stars," dodging coffee-flecked waiters and pushing empty chairs out of my way, until I reached them. The mathematician smiled at his wife (also a mathematician, I learned), and the surprisingly placid teenage girl, who looked a lot like her mother, sat in between. Their flight out was still a few hours away: over tea and toast, we talked.

We talked about the Four Color Theorem, which states

that all possible maps can be colored in such a way that no district or country touches another of the same color—using only (for instance) red, blue, green, and yellow. "At first sight it seems likely that the more complicated the map, the more colors will be required," writes Robin Wilson in his popular account of the puzzle's history, *Four Colors Suffice*, "but surprisingly this is not so." Redrawing a country's boundary lines, or imagining wholly alternative continent shapes, makes no difference whatsoever.

One aspect of the problem, in particular, had long intrigued me. After more than a century of fruitless endeavors to demonstrate the theorem conclusively, in 1976 a pair of mathematicians in the United States finally came up with a proof. Their solution, however, proved controversial because it relied in part on the calculations of a computer. Quite a few mathematicians refused to accept it: computers cannot do math!

"I actually met one of those guys who came up with the proof," my new friend recalled, "and we discussed how they had found just the right way to feed the data into the machine and get an answer back. It really was a smart result."

What did he and his wife think of the computer's role in mathematics? In answer to this general question they were more circumspect. The Four Color Theorem's proof, they admitted, was inelegant. No new ideas had been stimulated by its publication. Worse, its pages were almost unreadable. It lacked the intuitive unity, and beauty, of a great proof.

Beauty. How often have I heard mathematicians employ

this word! The best proofs, they tell me, possess "style." One can often surmise who authored the pages simply from the distinctive way that they were put together: the selection, organization, and interplay of ideas are as personal, and as particular, as a signature. And how much time might they spend on polishing their proofs. Superfluous expressions, out! Ambiguous terms, out! Yes, but it was worth all the trouble: well-written proofs could become "classics"—to be read and enjoyed by future generations of mathematicians.

"What time is it?" None of us was wearing a watch. We stopped a waiter and asked. "Already?" said the mathematician's wife when she heard his answer. They drained their cups, and dispersed their crumbs, and made shuffling sounds with their feet.

"Oh," said the mathematician, turning back to me, "I forgot: where did you say you were based again?" What with the history of the decimals, and the winding numerical vistas, and the painting of the entire globe with the colors of a single flag, the accidental features of our lives—where we lived, with whom, under what roof and color of sky—had been completely absent from our conversations.

I told him. "Paris," he echoed. "Why, we love Paris!"

France's capital has something of a one-sided reputation as the consummate city of artists. We know it as the city of Manet, of Rodin, of Berlioz; as the city of street singers and can-can dancers; as the city of Victor Hugo and of young Hemingway in *A Moveable Feast:* scribbling in a café corner, turning coffee and rum and the strictures of Gertrude Stein into stories. But Paris is also the city of mathematicians.

Its researchers, a thousand strong, make the *Fondation Sciences Mathématiques de Paris* (FSMP) the largest group of mathematicians in the world. About one hundred of the city's streets, squares, and boulevards are named after their predecessors. In the twentieth *arrondissement,* for example, one can walk the length of *rue* Evariste Galois, named after a nineteenth-century algebraist felled at the age of twenty by a dueler's bullet. On the opposite side of the Seine, in the fourteenth *arrondissement,* lies *rue* Sophie Germain, whose namesake introduced important ideas in the fields of prime numbers, acoustics, and elasticity before her death in 1831. According to her biographer Louis Bucciarelli, "She did not wish to meet others in the streets or houses of the day, but in the purer realm of ideas outside time, where person was indistinguishable from mind and distinctions depended only on qualities of intellect." A few minutes' walk away is Fermat's little road. There are also streets called Euler, and Leibniz, and Newton.

Among the letters waiting for me on my return to my adopted home was one from the city's *Fondation* Cartier. A museum for contemporary art, it had sent me a preview invitation to its upcoming exhibition "Mathematics: A Beautiful Elsewhere"—the first in Europe to showcase the work of major living mathematicians in collaboration with world-class artists. The timing seemed doubly auspicious: October 2011 happened to be the two-hundredth anniversary of Galois's birth.

The museum stands in the fourteenth *arrondissement* at the lower end of one of the long boulevards that diagram the city. It is an ostentatiously modern building, all shiny glass and geometric steel, bright and spacious, an example of "dematerialized" architecture. Reflected in the glass, scraggly trees denuded of their summer foliage appeared twice. I looked up at the symmetrical branches as I passed and entered.

Mathematics and contemporary art may seem to make an odd pair. Many people think of mathematics as something akin to pure logic, cold reckoning, soulless computation. But as the mathematician and educator Paul Lockhart has put it, "There is nothing as dreamy and poetic, nothing as radical, subversive, and psychedelic, as mathematics." The chilly analogies win out, Lockhart argues, because mathematics is misrepresented in our schools, with curricula that often favor dry, technical, and repetitive tasks over any emphasis on the "private, personal experience of being a struggling artist."

It was the mathematicians' artistic impulse, and inner struggle, that the exhibition's organizers intended both to communicate and celebrate. A white interior, zero-shaped, was the work of the filmmaker David Lynch. Walls usually reserved for frames and canvases lent their space to equations, light effects, and number displays. I walked through the rooms, now bare and silent, now colorful and stimulating, stopping here and there to take a closer look. I watched the other guests stand back and point and converse in low voices. Before a bright collage of sunrays and leopard spots, waves and peacock tails, and the underlying equations for each, fingers swayed and eyes widened. Another hall arrested visitors' feet around a lean aluminum sculpture, its curves reaching toward infinity.

But for me, the highlight of the exhibition took place in a darkened room downstairs. Here the visitors melted into twilight, rendered homogeneous in the darkness, sitting or standing in silence, all eyes, observing a large screen where a film shot in black and white was playing. A youngish face, screen big, was talking about his life as a mathematician. I pressed my back against the far wall and listened as he spoke of "fat triangles" and "lazy gases." Three or four minutes later, the film suddenly altered: the face gave way to another, wearing glasses. Four minutes after this, the face changed again: this time, a woman's voice began to speak about chance. In total, the film lasted thirty-two minutes—eight faces long. The men and women featured came from a wide range of mathematical subdisciplines—number theory, algebraic geometry, topology, probability—and spoke either in French, or English, or Russian (with subtitles), but their passion and wonder linked each personal testimony into a fascinating and involving whole.

Two of the testimonies, in particular, stood out. They reminded me of my conversations with the mathematicians in Mexico, and with those in other lands, and the feelings of kinship and excitement that these exchanges incited within me. During his four minutes, Alain Connes, a professor at the *Institut des Hautes Études Scientifiques,* described reality as being far more "subtle" than materialism would suggest. To understand our world we require analogy—the quintessentially human ability to make connections ("reflections" he called them, or "correspondences") between disparate things. The mathematician takes ideas that are valid in one area and transplants them into another, hoping that they will take and not be rejected by the recipient domain. The creator of "noncommutative geometry," Connes himself has applied geometrical ideas to quantum mechanics. Metaphors, he argued, are the essence of mathematical thought.

Sir Michael Atiyah, a former director of the Isaac Newton Institute for Mathematical Sciences in Cambridge, used his four minutes to speak about mathematical ideas "like visions, pictures before the eyes." As if painting a picture or dreaming up a scene in a novel, the mathematician creates and explores these visions using intuition and imagination. Atiyah's voice, soft and earnest, made attentive listeners of everyone in the room. Not a single cough or whisper intervened. Truth, he continued, is a goal of mathematics, though it can only ever be grasped partially, whereas beauty is immediate and personal and certain. "Beauty puts us on the right path."

The faces, old and young, smooth and hairy, square and oval, each had their say. Gradually, the room began to empty. Its intimate ambience slowly dissolved. I followed the last group of visitors up the stairs and out the building and not a word was exchanged. The night absorbed us.

I walked for a while, beside the river, with the night in my hair and in my pockets and on my clothes. The night, I know, is tender to the imagination; at this hour, throughout the city, artists sharpen pencils and dip brushes and tune guitars. Others, with their theorems and equations, revel just as much in the world's possibilities.

The world needs artists. Into words and pictures, notes and numbers, each transforms their portion of the night. A mathematician at his desk glimpses something hitherto invisible. He is about to turn darkness into light.

ACKNOWLEDGMENTS

I could not have written this book without the love and encouragement of my family and friends.

Special thanks to my partner, Jérôme Tabet.

To my parents, Jennifer and Kevin; my brothers, Lee, Steven, and Paul; and my sisters, Claire, Maria, Natasha, Anna-Marie, and Shelley.

Thanks also to Sigriður Kristinsdóttir and Hallgrimur Helgi Helgason, Laufey Bjarnadóttir, and Torfi Magnússon, Valgerður Benediktsdóttir and Grímur Björnsson, for teaching me how to count like a Viking.

To my most loyal North American readers, Margo and Linda Flah, Jean-Philippe Tabet, Valérie and Arnaud Salambier, and Claire Bertrand (and family!).

I am grateful to my literary agent, Andrew Lownie, and to Tracy Behar and Christina Rodriguez, my editors.

ABOUT THE AUTHOR

DANIEL TAMMET is the critically acclaimed author of the worldwide bestselling memoir *Born on a Blue Day,* and the international bestseller *Embracing the Wide Sky.* Tammet's exceptional abilities in mathematics and linguistics are combined with a unique capacity to communicate what it is like to be a savant. His idiosyncratic worldview gives us new perspectives on the universal questions of what it is to be human and how we make meaning in our lives. Tammet was born in London in 1979, the eldest of nine children. He lives in Paris.

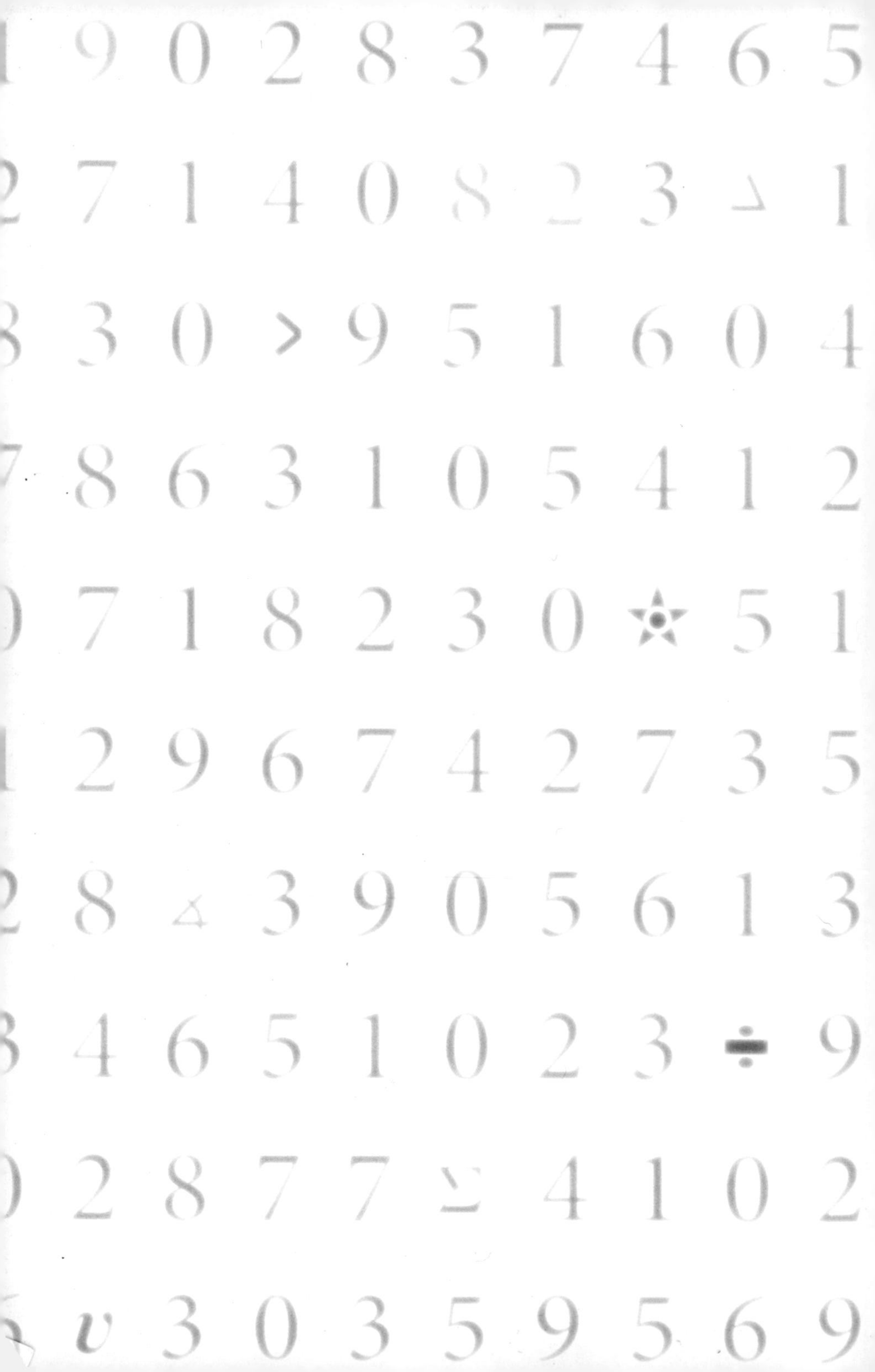